TABLE OF CONTENTS

Anthropogenic radionuclides

Materials Sciences

Education

Publications

Talks and Posters

Seminars

Theses

Collaborations

Visitors

EDITORIAL

In the age of social media, one could question whether or not it is worthwhile to provide a comprehensive overview of our research activities as a printed document. Indeed, we do get an overwhelming variety of information electronically, and also this document is available as portable file (pdf) on our homepage www.ams.ethz.ch. Nevertheless, we decided to continue our annual reporting in the old fashioned style. You may find a quiet moment, with a glass of wine or a good Whisky, take this report, and thumb through it. We are almost sure, you will find more than one study attracting your interests. As a tradition, the summaries will not go too deep into the details of the individual projects. It is instead our intention to stimulate your attraction for the great variety of research activities and that some studies may rise your interest to follow up with the related scientific literature. We hope you will enjoy browsing through this document, and that you give us a "like".

As a clear focus of our research, we are committed to progressing measurement technology. This of course requires a deep fundamental understanding of the physical processes behind the analytical methods and it is a special opportunity of our laboratory to have available the technical and operational background as well as the required support to design, construct, and set into operation new and unique instrumentation. In this respect, another breakthrough has been reached in 2017. Embedded into a research collaboration between ETHZ and Ionplus AG, we could finalize the first Multi-Isotope Low Energy AMS instrument (MILEA) which is not based on a traditional particle accelerator. In order to improve performance over our existing instruments, we implemented achromatic injection to achieve a better mass separation already at the low energy end, from which especially the heavy masses in the actinide region will benefit. Furthermore, and maybe more important, we included an electrostatic quadrupole triplet lens system in order to enable similar beam transport conditions for different ion species emerging from the acceleration stage in various charge states. Here, we got off the beaten tracks, exploited modern machining technologies, and created true hyperbolic electrodes forming the poles. We expect that this design minimizes aberrations while it increases the usable aperture without compromising image mapping quality. In such way, we consequently follow our mission to have at LIP the most advanced level of instrumental capabilities to make sure that not only our own applied research activities but also those of our partners and users can be conducted at the cutting edge of present day possibilities. We are proud that LIP managed to realize such a major project solely based on own, LIP internal resources and a substantial contribution from Ionplus according to their commitment within the collaboration agreement.

It turned out to be very efficient to have available two independent MICADAS systems for our radiocarbon program. While the new LIPMICADAS was primarily used for analyzing solid state graphite targets, the protoMICADAS covered all gaseous CO_2 analyses. In 2017, about 15'000 individual radiocarbon analyses were performed, and together with all the other nuclides we have measured over 20'000 samples, more than ever in LIP history. The broad range of applications covered by this report is a result of LIP being part of a strong and persisting network of collaborations with universities of Switzerland, Europe and overseas, national and international research and governmental organizations as well as commercial companies. We are most grateful for the confidence of every single user of our facility and for their ongoing support.

Thanks to the commitment of all staff members, LIP has provided extraordinary service to our internal and external users and also significantly contributed to the educational program in several departments of ETH. Without such an excellent scientific, technical and administrative staff, the long successful story of the Laboratory of Ion Beam Physics would not have been possible.

Hans-Arno Synal and Marcus Christl

THE LIP ACCELERATOR FACILITIES

Operation of the 6 MV TANDEM accelerator

Activities on the 0.5 MV TANDY system in 2017

Radiocarbon measurements on MICADAS in 2017

Assembly of the new 300 KV Multi-Isotope AMS

A new 1.7 MV Tandetron accelerator for LIP

Dust in the premises of LIP

OPERATION OF THE 6 MV TANDEM ACCELERATOR

Beam time statistics

Scientific and technical staff, Laboratory of Ion Beam Physics

The operation time in 2017 of the 6 MV tandem accelerator was divided into highly productive periods (more than 80% was used for actual measurements of samples) and downtime in summer due to construction work. The total operation time of 1203 hours is very similar to previous years (Fig. 1). About 37% of the time was dedicated to AMS and 57% to Material sciences, while only about 6% was used for maintenance activities.

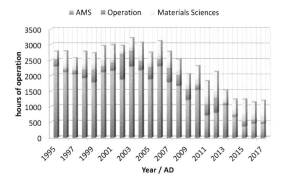

Fig. 1: *Time statistics of the TANDEM operation subdivided into AMS (blue), materials sciences and MeV-SIMS (green), and service and maintenance activities (red).*

We had three tank openings in 2017: in January the generating volt meter (GVM) was replaced with a new one from NEC eliminating the terminal voltage instabilities, which occurred in 2016. The second opening in May was necessary because the test with nylon sheaves was not successful and caused many sparks at higher terminal voltages. Failed bearings on the HE chain made a third tank opening inevitable. In addition, the 0° ion source was moved in order to gain space for the new 1.7 MV Tandetron accelerator (Fig. 2).

In 2016 we revived the gas-filled magnet (GFM) with a newly built large-acceptance gas ionization chamber. Initially designed for ^{26}Al measurements from AlO^-, we are now using it for ^{36}Cl. Using the GFM for ^{36}Cl makes the measurement more robust against ^{36}S interferences.

Despite the interruptions for the renovation work we had a significant increase of measured samples. The AMS activities focused on measurements of ^{36}Cl (now with the GFM) with 477 unknown samples. Most of the samples were geological applications and water samples.

Fig. 2: *Moving the 0° ion source to its new (original) position.*

The number of samples measured for materials science by the IBA techniques RBS, ERDA, and PIXE slightly increased by about 100 to 1370. All these analyses were performed for projects of curatorial partners and for commercial orders by industry. In addition to this another 1040 measurements were done for a master and a PhD project investigating secondary ion yields of large molecules induced by MeV cluster ions. The total beam time used for materials science increased by about 14% compared to 2016. This is largely due to the MeV-SIMS project.

ACTIVITIES ON THE 0.5 MV TANDY SYSTEM IN 2017

Beam time and sample statistics

Scientific and technical staff, Laboratory of Ion Beam Physics

In 2017, the ETH Zurich multi isotope system Tandy was running for more than 2800 hours and almost 3000 unknown AMS samples were analyzed for different radionuclides.

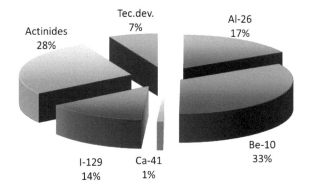

Fig. 1: *The old Pelletron chain which shows clear sign of heavy usage (metal contact bands of the wheels have cut into the pellets).*

After more than 33000 hours of total operation time (since 1998), the Pelletron chain including wheels and bearings had to be replaced (Fig. 1). During this major maintenance also one of the ADAM modules on the terminal was replaced, which is used to control the stripper gas regulation.

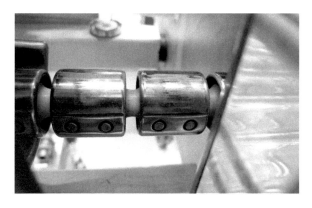

Fig. 2: *Relative distribution of the TANDY operation time for the different radionuclides and activities in 2017.*

The Tandy beam time in 2017 was mainly spent for routine AMS analyses of ^{10}Be, Actinides, ^{26}Al, ^{129}I, and ^{41}Ca for various projects and users (Fig. 2). The application projects include tracing artificial nuclides in the oceans, in situ dating, and ice core projects, as well as geological and environmental monitoring programs of our external users. Technical developments focused on the improvement of AMS setups for ^{26}Al and ^{10}Be measurements, by applying the absorber technique at low energies.

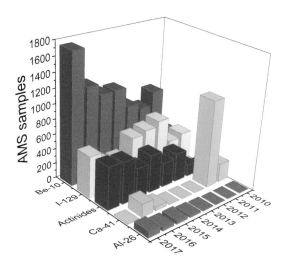

Fig. 3: *Number of AMS samples per nuclide measured since 2010.*

The annual number of AMS samples analyzed on the Tandy was never as high as in 2017 (Fig. 3). A major contribution to this were the large number of ^{10}Be samples analyzed to improve time resolution of the GRIP ice core ^{10}Be record (in collaboration with Univ. Lund) and the increased number of ^{129}I and actinide samples measured for our in-house oceanographic tracer program. In addition to that, several AMS analyses of ^{10}Be, ^{26}Al, ^{129}I, and actinides were performed for our about 20 external users who profit from the short turnaround times and individual support by the LIP AMS team.

RADIOCARBON MEASUREMENTS ON MICADAS IN 2017

Performance and sample statistics

Scientific and technical staff, Laboratory of Ion Beam Physics

For the second year, we run the radiocarbon samples on two MICADAS systems, after the installation of the GeoLipMicadas was installed at the beginning of 2016. The ProtoMICADAS was upgraded in 2017 to run on He stripping, which improved the measurement efficiency by nearly 30%. Additionally, the ProtoMICADAS obtained a new source head with a new extractor, where negative ions formed in the source are no longer decelerated. The GeoLipMICADAS now operates with a pulsing of -3/0/+3 kV instead of 0/+3/+6 kV for ^{14}C, ^{13}C and ^{12}C, respectively. From this change, we expect less stress on the foil insulating the magnet chamber.

17500 samples were measured in 2017, which is significantly higher than in 2016. The GeoLipMICADAS nearly continuously ran on graphite samples, while the ProtoMICADAS was primarily used for gas measurements. During some weekends graphite samples were also measured on the ProtoMICADAS.

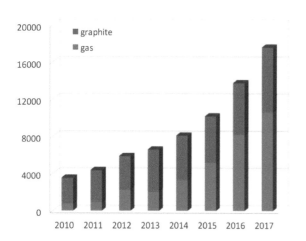

Fig. 1: *The amount of analyzed radiocarbon samples steadily increased over the years. 17 500 sample targets were measured in 2017.*

For both, gas and solid samples, we measured >25% more samples than in 2016 (Fig. 1).

The increased number of measured gas samples is primarily due to the 1700 biomedical samples that were analyzed in scientific collaboration with Novartis (Fig. 2) [1].

More than 1000 graphite samples were measured for the SNF project that aims at extending the IntCal calibration curve throughout the Younger Dryas [2]. Another 500 samples were analyzed for a project to reconstruct the atmospheric ^{14}C concentrations with tree-rings in annual resolution. This resulted in a large increase of measured graphite samples of more than 25%.

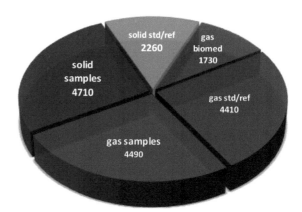

Fig. 2: *Samples measured on the two LIP MICADAS systems in 2017. Graphite samples and standards are shown in blue. Red and grey indicate gas samples.*

The amount of samples measured for our partner institutions stayed with 3200 nearly constant.

[1]   D. De Maria et al., LIP Annual Report (2017) 32
[2]   A. Sookdeo et al., LIP Annual Report (2017) 39

ASSEMBLY OF THE NEW 300 KV MULTI-ISOTOPE AMS

Construction steps and the current project status

S. Maxeiner, M. Christl, A. Müller, M. Suter, H.-A. Synal, C. Vockenhuber, R. Pfenninger, S. Bühlmann, K. Seidler, R. Gruber, P. Vogel, B. Käsermann, J. Thut, B. Fehst

A concluding proof-of-principle experiment was performed in 2015/2016 for a compact 300 kV multi-isotope AMS system [1]. Soon after, the design and assembly of the new Multi Isotope Low Energy Accelerator (MILEA) began. Construction started in the LIP measurement hall in June 2017 with the placement of the support frame, followed by the delivery of the three magnets in the following month.

Fig. 1: The new quadrupole focusing system (view along the beam axis) [2]. Its main purpose is ensuring optimal ion optics for all charge states and isotopes.

Assembly and alignment of the beam line with the remaining elements, such as the ion source, injector- & analyzer ESA, analyzing chambers and the new quadrupole system (Fig. 1, [2]) was done in collaboration with IonPlus AG. In parallel, electronic components, water and compressed air supplies were implemented and excellent vacuum could be reached with MILEA in December 2017. In the same year, the maximum required high voltage for all elements except for the accelerator was tested and reached without encountering any issues.

Fig. 2: The MILEA facility in January 2018.

Performance tests started in January 2018 with the successful measurement of carbon ions at 200 kV acceleration voltage. Measurement of further isotopes and the herewith involved need for higher acceleration voltages up to the maximum of 300 kV are the next steps before the system can transition into routine operation and serve as a research platform. New know-how needs to be gained especially with the handling and performance of the new quadrupole system, the novel type Faraday cups [3] and the measurement of isotopes other than carbon using MICADAS-type target holders.

[1] S. Maxeiner et al., LIP Annual Report (2016) 14
[2] S. Maxeiner et al., LIP Annual Report (2017) 19
[3] M. Christl et al., LIP Annual Report (2016) 13

A NEW 1.7 MV TANDETRON ACCELERATOR FOR LIP

Moving an accelerator from La Chaux de Fonds to Zurich

E. Guibert, S. Bühlmann, M. Döbeli, R. Gruber, H.-A. Synal, P. Vogel, C. Vockenhuber

In 2017, Haute Ecole Arc Ingénierie in La Chaux-de-Fonds, NE, decided to abandon its activities in ion beam analysis and to stop the operation of the 1.7 MV Tandetron accelerator facility, which is a dedicated system from the company High Voltage Eng. Europe (HVEE) for material sciences that was installed in 2008. In an agreement between the État of Neuchâtel, Haute Ecole Arc and ETH Zurich, the accelerator facility and most of its components were handed over to LIP.

Fig. 1: *Removing the acceleration column out of the tank for complete disassembly at La Chaux-de-Fonds.*

The dismantling started in June 2017. All beam line components were carefully packed and the accelerator itself completely disassembled with the help of a technician of HVEE (Fig. 1). The transport took place in October, in total three 40 t truckloads were necessary to transport all components to Zurich (Fig. 2). At LIP we removed the Super-SIMS ion source and part of the 0° ion source to have enough space for the new accelerator in the accelerator hall behind the ion sources of the 6 MV Tandem. Additionally, the opening of the thick concrete wall was enlarged and two concrete shielding doors dismantled.

Fig. 2: *The main accelerator tank is arriving at ETH Zurich.*

In December 2017, we started with the installation of the Tandetron. A long transfer line will bring the beam behind the 6 MV Tandem to the new 90° magnet. From there the beam will basically follow the old beam lines, although the beam line height of the new accelerator is only 1.2 m. In order to do that, the old magnets and beam lines have to be removed once the ion sources and the accelerator are working. The experimental stations will be a mix of existing chambers (for RBS and ERDA) and equipment from La Chaux-de-Fonds, including a high resolution RBS spectrometer and an Oxford Micro-Beam system. It is planned to have the facility operational in 2018.

DUST IN THE PREMISES OF LIP

Making space for a new accelerator and future activities

Scientific and technical staff, Laboratory of Ion Beam Physics

In summer 2017, we started several renovation activities in the measurement and Tandem hall as well as in the control room.

For the new 1.7 MV Tandetron accelerator, space was acquired behind the ion sources of the 6 MV Tandem by removing the Super-SIMS ion source, the coupling ESA and the HASY lens and by moving the 0° ion source forward to its original position. On the HE side, the wall made of concrete shielding blocks and the concrete door were removed (Fig. 1).

Fig. 2: *A temporary housing was installed to control the dust from the cut-through.*

Fig. 1: *Removing the concrete shielding blocks in the Tandem hall, the concrete door at the left was also dismantled and removed.*

The µCADAS was removed from the measurement hall and the TANDY control console was relocated in order to make space for the MILEA facility between the ProtoMICADAS and the TANDY facility. The opening in the concrete wall between the Tandem and the measurement hall was enlarged (Fig. 2).

In the control room we removed the concrete shielding door in order to have more general space (Fig. 3). In addition, many old and now unused cables and electronic racks of the 6 MV Tandem were removed.

Fig. 3: *Dismantling the concrete door with heavy equipment in the control room.*

All these activities happened during the normal operation and measurement activities at LIP with one exception: in July and August the beam at the 6 MV Tandem had to be shut down for a few weeks for the cut-through.

INSTRUMENTAL DEVELOPMENTS

^{36}Cl measurements with gas-filled magnet

Towards efficient ^{10}Be measurements at TANDY

Improvements at the MeV-SIMS setup CHIMP

Quadrupole considerations

Redesigning the ETH current integrators

Double trap interface

Radiocarbon measurement of atmospheric CO_2

Status of the ETH *in situ* ^{14}C extraction line

^{36}Cl MEASUREMENTS WITH GAS-FILLED MAGNET

Additional ^{36}S suppression for improved reliability of ^{36}Cl AMS

C. Vockenhuber, K.-U. Miltenberger, M. Suter, H.-A. Synal

^{36}Cl measurements have a long history at LIP: already in the 1980s we showed that ^{36}Cl measurements are possible at energies available from 6 MV accelerators (48 MeV) [1]. A strong program has developed over the decades, which was driven by a large number of measurements of ice and water samples, as well as geological applications.

Based on ^{36}Cl-^{36}S separation in the gas ionization chamber (GIC) we achieve a high overall efficiency. However, for samples with high S content counting rates can get higher than a few 1000 cts/s and are therefore problematic for the GIC.

With the gas-filled magnet (GFM) we can now separate ^{36}Cl from ^{36}S by a factor of ~300 and reduce the total counting rates in the GIC, while keeping the efficiency high (the transmission through the GFM is ~75%). Identification of ^{36}Cl is done in the GIC, which provides another ^{36}S suppression factor of ~300, resulting in a total ^{36}S suppression of 10^5, which is similar to the GIC-only method.

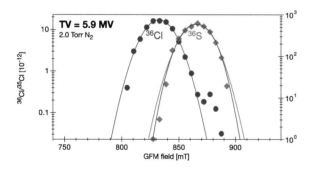

Fig. 1: *Separation of ^{36}Cl and ^{36}S in the GFM filled with 2 Torr N_2 at 47.2 MeV (TV=5.9 MV).*

To find optimal operating conditions (Fig. 1) for the GFM and the newly build large acceptance GIC, we performed systematic investigations [2].

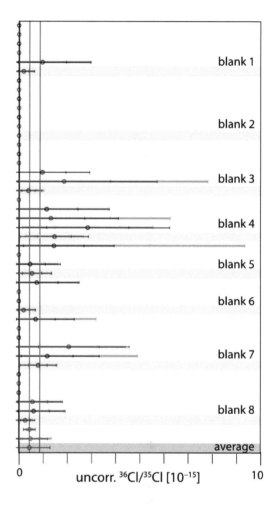

Fig. 2: *User blank measurements over the period of one week (the shown values are not normalized), resulting in a final normalized average of ^{36}Cl/^{35}Cl = (0.8±0.7) x 10^{-15}.*

In 2017 we switched to the GFM method for routine measurements of ^{36}Cl and measured about 500 samples. In general, the performance is similar to the GIC-only method, however, we are less sensitive to S content in the sample. Blanks are now consistently measured around 10^{-15} (Fig. 2).

[1]   H.-A. Synal et al., NIMB 29 (1987) 146

[2]   C. Vockenhuber et al., submitted to NIMB

TOWARDS EFFICIENT ^{10}Be MEASUREMENTS AT TANDY

Investigation of the ^{1}H background with the absorber setup

M. Bryner, M. Christl, C. Vockenhuber

The efficiency of ^{10}Be measurements at the Tandy with the current degrader method has an overall transmission of only about 13% [1]. Alternatively, the interfering ^{10}B can be removed in an absorber setup. However, the residual energy of the ^{10}Be in the detector (when the ^{10}B is completely stopped in the absorber) is very low (about 100 keV), making a clear identification challenging. Despite using the 2+ charge state with a lower transmission through the accelerator of 25%, the overall transmission could be increased to more than 20% with the absorber setup. However, with the new setup, a background appears at the level of 10^{-14} (Fig. 1), which turns out to be ^{1}H coming from the absorber foil or the absorber gas due to scattering with incoming ^{10}B atoms. The absorber setup for ^{10}Be will only be competitive, if this background can be reduced or separated.

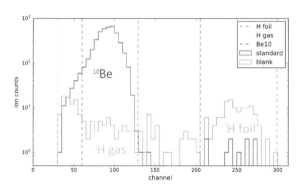

Fig. 1: *Residual energy spectrum of a standard (blue) and a blank (green) sample.*

In order to reduce this background, various methods have been investigated. Simulations were run using SRIM; different absorber gases and window holders were tested and most notably, measurements have been made with different detector pressures. The idea was to measure not only the residual energy of the ions (Fig. 1), but also the energy loss over distance using two detector anodes. Since ^{1}H loses much less energy in the detector gas than ^{10}Be, separation should generally be possible.

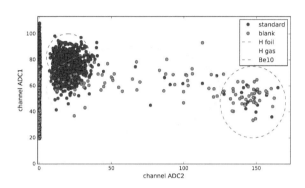

Fig. 2: *2D spectrum of a standard and a blank sample, due to the low ion energies the regions for ^{10}Be and ^{1}H from the gas overlap. Events on the y-axis do not reach the second anode.*

The detector pressure was adjusted so that only ^{10}Be reaches the first anode, while the ^{1}H atoms travel further and also deposit energy on the second anode. The second anode, hence, can be used as a veto. It was found that even for very low detector pressures, a fraction of about 10% of the ^{1}H never reaches the second anode. This can be explained by the geometry of the detector. At even lower detector pressures the ^{10}Be ions reach the second anode allowing a separation in a 2D spectrum (Fig. 2). However, at these low pressures the ^{1}H ions are not fully stopped inside the detector anymore and the regions for ^{1}H from the gas and ^{10}Be in the spectrum begin to overlap.

Further improvements of the current absorber setup are necessary to reduce the ^{1}H interference so that ^{10}Be measurements at the Tandy can benefit from the higher overall transmission.

[1]   A.M. Müller et al., NIMB 268 (2010) 2801

IMPROVEMENTS AT THE MEV-SIMS SETUP CHIMP

Modified extraction and electron shielding boost ToF performance

K.-U. Miltenberger, M. Döbeli, H.-A. Synal

In parallel to performing imaging, resolution and yield measurements with the MeV-SIMS setup CHIMP [1,2], it was continuously improved over the last year. After installation of an additional high voltage biased shielding for suppression of electrons originating at the collimating slits and capillary entrance, which improved the background considerably, the extraction for the positive ion ToF spectrometer was revisited.

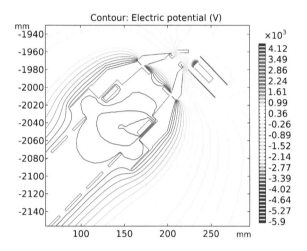

Fig. 1: *Simulation of the modified electric extraction field around the ToF extraction cone, sample and channeltron electron detector.*

In August, sparking between the two sections of the extraction cone, at that time held at a potential difference of 3.8 kV, was detected and consequently an additional PTFE insulation was installed. At the same time, the voltage supplied to the Micro Channel Plate (MCP) front was separated from the liner voltage to obtain an additional free parameter for tuning of the spectrometer. Optimum performance is now obtained when the second cone and liner are held at a potential of -4 kV (only 2 kV lower than the first cone element) resulting in particle flight times that are increased by ~15 %. The MCP front voltage could be slightly increased to -6 kV such that the ions are accelerated a second time

just in front of the microchannel-plate detector. A simulation of the resulting electric field configuration is shown in Fig. 1.

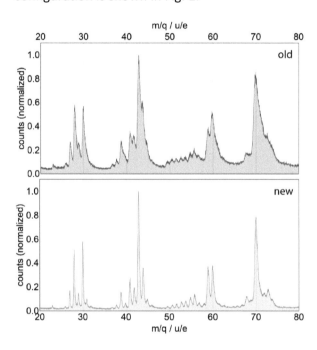

Fig. 2: *Comparison of mass spectra (selected mass region m/q = 20–80 u/e) measured on an Arginine sample with the old and new ToF settings (using 28 MeV $^{197}Au^{7+}$ resp. 15 MeV $^{127}I^{4+}$ primary ion beams).*

As shown in Fig. 2 the new configuration significantly increased the mass resolution of the ToF spectrometer from $m/\Delta m \approx 45$ to $m/\Delta m > 100$.

[1] M. Schulte-Borchers et al., NIMB 380 (2016) 94

[2] K.-U. Miltenberger et al., NIMB 412 (2017) 185

QUADRUPOLE CONSIDERATIONS

True hyperbolic shape or circular approximation?

S. Maxeiner, H.-A. Synal

In an ion accelerator, as for example an AMS system, the ion beam needs to be focused several times along its path from the source into the detector by lenses. This ensures high transmission through the system and allows to define focal points, which enable efficient and sensitive measurements. The tandem acceleration itself is an electrostatic lens with physical properties that strongly depend on the ion's charge state and energy. In a multi-isotope AMS this can become a limiting factor as each isotope is focused differently after stripping while the spectrometer is optimized for only one isotope.

One of the key features of the new multi-isotope system built at LIP [1] is the electrostatic quadrupole triplet lens right after the tandem acceleration. It ensures equal ion optics for all isotopes and charge states after stripping. These kinds of lenses consist of four parallel metal electrodes of hyperbolic shape (Fig. 1, blue). An alternative shape, which is often favored instead are circular rods (Fig. 1, yellow), because they are much easier to manufacture and align.

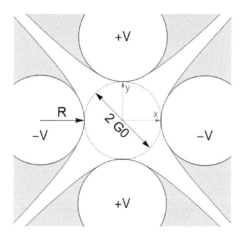

Fig. 1: View along the beam axis: cross section of hyperbolic (blue) or circular (yellow) electrodes with radius R. The electrodes are charged to voltages ±V.

The focusing fields along the x axis generated by different shapes are shown in Fig. 2. With circular rods, deviations from the perfect field become larger with increasing distance x from the central beam axis and introduce lens aberrations, which lead to degraded performance of the spectrometer.

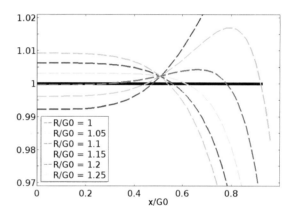

Fig. 2: Focusing field generated by electrodes of different radii R normalized to the perfect quadrupole field with hyperbolic shapes (black line).

The usable area of a quadrupole lens therefore has a radius which is smaller than the bore radius G0 in case of circular electrodes. Increasing the bore radius helps in this case but increases the required electrode voltages proportionally to $(G0)^2$. Electric breakthrough in vacuum becomes more likely and the lens grows in size. For the new facility we therefore decided to go for true hyperbolic shapes (a big thank you to the ETH workshop!) to utilize the full bore radius G0 for keeping voltages small, the unit compact and beam quality as high as possible.

[1] S. Maxeiner et al., LIP Annual Report (2017) 11

REDESIGNING THE ETH CURRENT INTEGRATORS

Precise current measurements at the pA level

M. Christl, A. Müller, P. Eberhard, S. Maxeiner, H.-A. Synal

For some AMS applications such as the analysis of ^{236}U in a few liters of seawater or the carrier-free measurement of ^{10}Be not only the detection of the rare isotope is challenging. The amount of the stable or abundant isotope in such samples is at the level of a few micro grams or even less. On the compact ETH AMS system Tandy these samples typically produce positive ion currents in the pA to nA range on the high energy side.

In contrast to conventional mass spectrometry, where the ion currents are typically measured in DC mode, an AMS current integrator has to work with pulsed ion beams and therefore collect and integrate the charge of a rather small number of ions entering the Faraday Cup within about a millisecond.

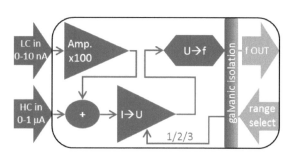

Fig. 1: *Block diagram of the redesigned ETH current integrator (I: current, U: voltage, f: frequency, LC/HC: low/high current).*

In collaboration with IonPlus, the existing ETH current integrator design was optimized for low noise and low level current measurements and it was additionally equipped with a software controlled range selector (Fig. 1). In addition to the normal (HC) input the new ETH VII integrator is additionally equipped with a low current input (LC). The user has to select between the LC and HC mode by connecting the Faraday cup to either input. For each operational mode a three level measurement range selection is available (Tab. 1), which is controlled by the user via the measurement software.

mode	range	t_{int} [μs]	I_{max} [nA]	q_{max} [pC]
HC	1	510	1000	510
	2	1700	300	
	3	5100	100	
LC	1	510	10	5.1
	2	1700	3	
	3	5100	1	

Tab. 1: *Operational modes and parameters of the ETH VII current integrator: integration time, full range, and max. av. charge per cycle.*

Fig. 2: *Picture of the new ETH VII integrator.*

The new design was realized in SMD technology (Fig. 2) to allow fast and easy assembly. Without hardware changes, the sensitivity of the ETH VII integrator can be adjusted by one order of magnitude with the range select controller. The additional equipment with the LC input that includes low noise current amplification allows that the full range of the current integrator can be varied by a factor of 1000 (Tab. 1).

First tests document that the new ETH VII integrator operated in the most sensitive mode has a precision of less (better) than 10^{-3} (of full range), which generally allows current measurements from the (sub) pA level to 1 μA.

DOUBLE TRAP INTERFACE

Design process and development of a new gas interface

D. De Maria, S. Fahrni[1], L. Wacker, H.-A. Synal

In the last years the development of compact AMS facilities like the 200 kV MICADAS has led to a broad application of this technology in many research fields. However, the low sample throughput is still a limiting factor for this technology. In particular, in the case of biomedical applications, where ^{14}C has a great potential as microtracer for pharmacokinetic studies, the need to measure a large number of samples in a short period of time limits an AMS approach to these studies. Therefore, our aim is to develop a new gas handling system to achieve a faster measurement process for applications where low precision but higher sample throughput is required.

Fig. 1: *Picture of the developed prototype of the double trap interface for the handling of gas samples.*

In the current Gas Interface System (GIS), used for routine gas measurements with MICADAS at LIP, the CO_2 deriving from the combustion of samples with an Elemental Analyzer (EA, Elementar GmbH, Germany) is collected using a trap containing a molecular sieve material (zeolite X13). The gas is then released into a syringe, where it is diluted and fed into the AMS system at a constant carbon mass flow. To avoid cross-contamination effects, an extensive cleaning procedure of both trap and syringe is required after each sample. Especially the trap cleaning is a time consuming process.

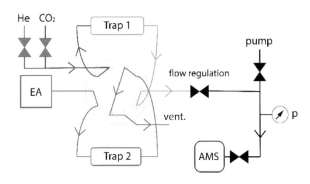

Fig. 2: *Schema of the double trap interface coupling the EA to the AMS system for gas measurements.*

The new design features two traps working in an alternating way. The gaseous CO_2 released from the traps is flushed directly into the AMS system, skipping the use of the syringe. In order to keep a constant mass flow into the source, the feeding pressure is regulated by a system of valves. The two optimized traps contain less zeolite material (X13), such that the adsorbed carbon quantity is constant regardless of the sample size. First tests of the new setup will be performed to investigate cross-contamination effects between consecutive samples and to determine its overall performance.

---

[1] *Ionplus AG, Dietikon*

RADIOCARBON MEASUREMENT OF ATMOSPHERIC CO$_2$

Direct graphitization of CO$_2$ from air samples with AGE for ^{14}C analysis

P. Gautschi, L. Wacker

Radiocarbon measurements of atmospheric carbon dioxide (CO$_2$) bear valuable information about Earth's carbon cycle. In particular the combination of ^{14}C analysis with the actual CO$_2$ concentration measurements allows for the quantification of locally emitted fossil fuel material [1]. However, the low atmospheric CO$_2$ concentration (approx. 400 ppm) makes the sample preparation difficult and time-consuming. Current techniques involve either the use of liquid nitrogen (LN$_2$) [2] or sodium hydroxide (NaOH) [3] for CO$_2$ separation. Both approaches are labor intensive and require careful sample handling.

A new technique for simple and fast air sample preparation using the existing Automated Graphitization Equipment (AGE) [4] was developed, hereby skipping the step involving LN$_2$ or NaOH. The device consists of a small membrane pump, which transfers 5 L of dried air into the AGE where CO$_2$ is directly separated from other atmospheric gases by the integrated molecular sieve cooled down to -4 °C. Then the sample's CO$_2$ content is graphitized and measured by AMS on the MICADAS [5]. A schematic diagram of the new setup is shown in Fig. 1.

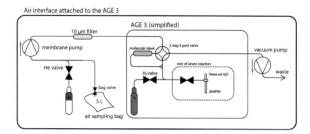

Fig. 1: *Schematic diagram of the new setup for air sample preparation.*

A high sampling yield of >90% allows to obtain 1 mg carbon while sampling only 5 L of air. To prove the applicability and simplicity of the new technique, more than 90 individual atmospheric air samples have been analyzed in the time span of four months. This correspond to roughly 1-2 samples per week, each with 2-7 duplicates. Additionally, more than 85 reference air samples and 66 blank air samples from pressurized gas cylinders were measured, to check for possible drifts and contamination.

Overall good performance was found for the novel setup with a precision of 2.1‰ for modern samples, a blank level of 0.0016 F^{14}C and cross contamination on the order of 1.1‰, while at the same time minimal user input (changing of sample bags) was required.

Fig. 2: *Picture of the open ALF system.*

Further steps towards full automation and batch preparation are already done. The design of the Air Loading Facility (ALF) will allow for up to seven individual air samples to be loaded consecutively into the AGE where they are subsequently graphitized simultaneously. Fig. 2 shows a photographic representation of the ALF with its top cover removed.

[1]   I. Levin et al., Radiocarbon 31 (1989) 431
[2]   T. A. Berhanu et al., ACP 17 (2017) 10753
[3]   I. Levin et al., Radiocarbon 22 (1980) 379
[4]   L. Wacker et al., NIMB 268 (2010) 931
[5]   H.-A. Synal et al., NIMB 259 (2007) 7

STATUS OF THE ETH *IN SITU* ^{14}C EXTRACTION LINE

Performance of the new ETH extraction system

K. Hippe, M. Lupker[1], L. Wacker

Applied *in situ* ^{14}C extraction methods and protocols have largely been developed during the 1990s. However, *in situ* ^{14}C extraction from terrestrial rocks is still a technically challenging process that is not yet widely employed. In 2016/2017, a new *in situ* ^{14}C extraction system has been built at LIP (Fig. 1) to improve the existing methods and provide more reliable extraction techniques.

At the ETH system, *in situ* ^{14}C extraction from quartz is achieved by high-T diffusion. Major steps for *in situ* ^{14}C analysis are: (i) weighing and loading of clean quartz (~2-5 g) into the furnace, (ii) sample pre-heating at 500°C (2 h) to remove atmospheric ^{14}C adsorbed to the crystal surfaces, (iii) release of the *in situ* component through heating to 1650°C (3 h), (iv) oxidation of all carbon species into CO_2, (v) gas purification by passage through hot Cu/Ag, a chemical water trap and by cryogenic sublimation, (vi) measurement of the extracted amount of CO_2 prior to sample transfer to the MICADAS AMS system.

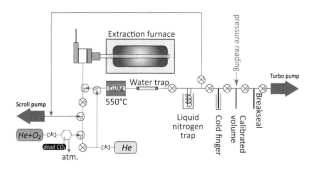

Fig. 1: *Sketch of the new extraction system.*

The new extraction system is largely automated and uses Helium for gas transport and cleaning/flushing. Only the final gas cleaning steps and the pressure measurement are done manually at the high-vacuum part at the end of the extraction line. Automation together with a significant simplification of the gas cleaning

procedure allowed reducing the attendance time from an entire working day to about 1.5 h per day. Currently, complete extraction of one sample is achieved within 12 hours. First system blanks (without heating the extraction furnace) are on the order of 2-3 x 10^4 at ^{14}C. Full procedural blanks (1650°C for 2-3 h) are highest immediately after sample extraction and appear to record insufficient cleaning of the system by He flushing. After additional flushing with He+O_2, procedural blanks decrease and stabilize at ~4-5·10^4 ^{14}C atoms (Fig. 2), which is comparable to blanks measured at the first ETH extraction line.

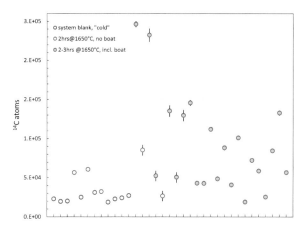

Fig. 2: *First procedural blank data for the new in situ ^{14}C extraction system. The light red bar indicates the blank level achieved after thorough system flushing.*

Yield test using the CRONUS-A quartz standard are currently ongoing and the extraction of geologic samples is envisioned to start in early 2018.

---

[1] *Geology, ETHZ*

RADIOCARBON

Terrestrial organic matter export

Carbon stability on a Hawaiian climate gradient

Riverine fire-derived black carbon (BC)

Riverine carbon export in the central Himalaya

Riverine Luzon Island organic carbon

Tracing deep sea organic carbon transport

The petrified terrigenous carbon trail of Taiwan

What on earth have we been burning?

ACTIVITIES IN THE ^{14}C LABORATORY IN 2017

Samples prepared for ^{14}C analysis and respective turnover time

I. Hajdas, S. Bollhalder, L. Hendriks, R. Hopkins, M. Maurer, M.B. Röttig, A. Sookdeo, A. Synal, L. Wacker, C. Welte, C. Yeman

Over 4000 samples were pre-treated and prepared by the ETH radiocarbon laboratory (Fig. 1, Tab. 1). Often multiple preparations (treatment and fractions) as well as multiple targets (graphite as well as gas) are required as reflected in the number of total analysis performed in 2017 [1].

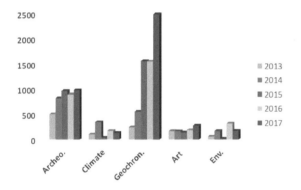

Fig. 1: *Number of samples (objects) analysed for various research disciplines during the last four years.*

The increase in samples analyzed for 'Geochronology' is due to samples prepared as a part of an intern research project (Tab. 1) dedicated to the extension of the calibration curve, where numerous tree ring samples were analyzed [2].

Research	Total	Internal
Archaeology	979	20
Past Climate	140	8
Geochronology	2487	2200
Art	281	129
Environment	182	150
Total	4069	2507

Tab. 1: *Number of samples analysed in 2017 for various applications. Column 'Internal' is the number of samples supported by the laboratory for PhD, master or term theses.*

This is also reflected in the portion of wood (60%) as material prepared in the laboratory. It is followed by charcoal (10%), which is the most common material analyzed for archeology. Other material includes bones, textiles, paper and paint. An ETH internal research project dedicated to studies of paintings involves analysis of multiple fractions [3] as it is the case for mortar samples [4].

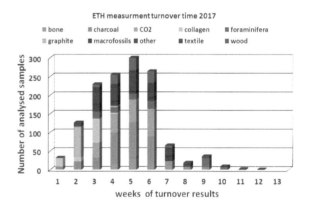

Fig. 2: *Turnover time for samples prepared at the ETH radiocarbon laboratory in 2017.*

The analysis turnover time is important when samples are submitted for ^{14}C analysis. Dependent on the type of samples, i.e. complexity of preparation, the overall analysis time might differ (Fig. 2). Results for small batches (1-5) or samples ready to be graphitized such as CO_2, collagen, or foraminifera can be returned in 1-2 weeks. The turnover time of 8-12 weeks is typically due to low carbon content or poor preservation of the sample. Such cases require re-sampling, additional treatment, or gas measurements.

[1]   L. Wacker et al., LIP Annual Report (2017) 10
[2]   A. Sookdeo et al., LIP Annual Report (2017) 39
[3]   L. Hendriks et al., Radiocarbon 60 (2017) 207
[4]   I. Hajdas et al., Radiocarbon 59 (2017) 1845

SIMPLE CO_2 EXTRACTION FROM SEAWATER

A new extraction method for the analysis of ^{14}C in salty water samples

A.-M. Wefing, N. Casacuberta, S. Bollhalder, L. Wacker

A new method for the extraction of CO_2 from seawater samples has been developed and tested, with the aim of measuring ^{14}C concentrations in water profiles in the ocean.

The existing extraction and graphitization line used for carbonate and air samples at LIP was slightly modified for the preparation of water samples. Briefly, water samples are collected in 120 ml glass bottles, sealed with a rubber septum, and poisoned with $HgCl_2$ to destroy any organic carbon. For the extraction of CO_2, about 60 ml of seawater are needed. The collection of 120 ml therefore allows for a repetition of the measurement. 60 ml of the sample are transferred to a second, clean bottle, that has been previously sealed and flushed with He. By adding 1 ml of phosphoric acid, the carbon contained in the sample is converted to CO_2, which is driven out of the water phase by heating the samples to about 60°C overnight. For the CO_2 extraction from the sample, He is flushed through the water with a flow rate of about 180 ml/min using a double needle.

The extracted CO_2 is first passed through a sicapent water trap and then through a heated glass tube (550°C) filled with Cu, before it is trapped on zeolite in the AGE graphitization system [1] (Fig. 1). Finally, CO_2 is thermally released to one of the seven reactors in the system and graphitized on iron powder.

Tests were conducted with and without the heated Cu tube. If the tube was not used, graphitization times for water samples containing salt (seawater or salty lake water) exceeded those for freshwater samples by far or the graphitization did not work at all (Fig. 2a).

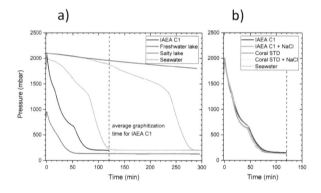

Fig. 2: Graphitization times for different types of samples, without (a) and with (b) the use of a heated Cu tube during CO_2 extraction.

The method was validated by repeated measurements of a coral standard dissolved in MilliQ water. Within uncertainty, the measured $F^{14}C$ values match the literature value for this standard [2].

[1] L. Wacker et al., NIMB 268 (2010) 931

[2] E.N. Hinger et al., Radiocarbon 52 (2010) 69

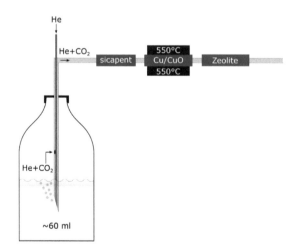

Fig. 1: Sealed bottle with double needle to extract CO_2.

DATA REDUCTION FOR LA-AMS

Optimizing spatial resolution by offline data reduction

C. Yeman, M. Christl, L. Wacker, B. Hattendorf[1], H.-A. Synal

In Laser Ablation AMS a pulsed laser is focused onto a sample's surface, producing CO and CO_2 that is introduced into the gas ion source for radiocarbon (^{14}C) analysis [1]. It is used for carbonate climate archives. As the laser crosses the growth layers on the sample the ^{14}C content of the ablated material and therefore of the produced gas changes. At the AMS the discrete counting time - a single cycle - is set to 10 seconds for LA AMS. The continuous change in ^{14}C of the gas is detected in the single cycles, providing a quasi-continuous temporal ^{14}C profile. Knowing the starting and ending point on the sample and the velocity with which the sample is moving under the laser, the temporal ^{14}C profile can be assigned a location on the sample.

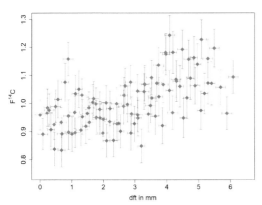

Fig. 1: *Single cycle data of three scans corresponding to the different colours.*

An example for a measurement is shown in Figure 1. The single cycle ^{14}C data of three laser scans are aligned with the position of the sample. The scans were run on top of each other analyzing the same area on the sample for ^{14}C. In an offline data reduction we are able to choose the area over which the data is integrated and therefore the spatial resolution. In order to do that every single cycle data point is assumed to be the weighted mean of ten subsamples with the same ^{14}C content therefore

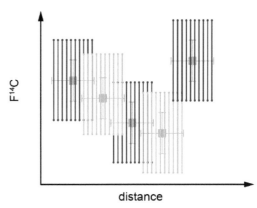

Fig. 2: *Single cycle data points divided in subsamples.*

$F^{14}C_{sub} = F^{14}C_{meas}$ and $\sigma_{sub} = \sigma_{meas} \cdot \sqrt{10}$. In a next step all subsamples of every scan covering an area interval are used to calculate the weighted mean. The intervals are chosen to optimize spatial resolution and precision. Applying these steps to the data from Fig.1 for two different spatial resolutions result in the ^{14}C profiles shown in Fig 3. A higher spatial resolution leads to less precision and vice versa.

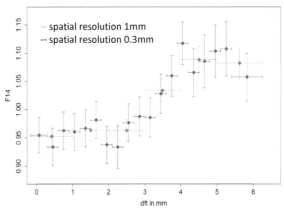

Fig. 3: *Weighted mean of the three scans for two different spatial resolutions.*

[1]   C. Welte et al., Anal. Chem. 88 (2016) 8570

[1] *Laboratory of Inorganic Chemistry, ETHZ*

^{14}C ANALYSIS OF ARCTICA ISLANDICA WITH LA-AMS

Using ^{14}C to trace the bomb peak in the North Sea

C. Yeman, R. Witbaard[1], M. Christl, L. Wacker, B. Hattendorf[2], H.-A. Synal

Arctica islandica is a bivalve mollusk that is found on the continental shelves of the North Atlantic Ocean. It is a long-lived species (up to 500 years) that grows in annual layers. This makes it an ideal proxy archive for the marine environment.

Here, we use laser ablation (LA) AMS [1] to analyze a specimen for radiocarbon (^{14}C) that has been captured in 1988 and has 66 yearly growth increments.

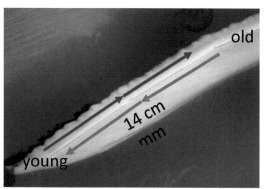

Fig. 1: *Section of the shell with the laser tracks.*

Fig.1 shows a section of the shell with the laser tracks. The arrows indicate the direction of the scans, crossing the growth layers and the distance covered during the measurement on one target. For visualization of the growth layers and their alignment with the ^{14}C data, an acetate peel of the section with the laser tracks was made as shown in Fig.2.

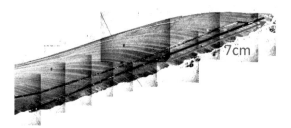

Fig. 2: *Acetate peel of the shell showing growth layers and laser tracks.*

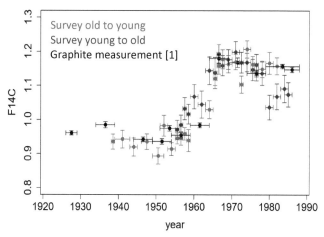

Fig. 3: ^{14}C *data of the survey scans and the graphite measurement done previously [2].*

The LA-AMS results, shown in Fig.3 are in agreement with the ^{14}C-AMS analysis of micro-drilled and graphitized samples that had been done previously [2]. The total LA-AMS measurement time was less than 2 hours and revealed the complete ^{14}C bomb peak in the shell. Compared to the samples extracted by drilling the ultra-fast LA AMS technique gives a higher spatial resolution while at the same time using less material. These advantages can now be exploited for the application in a broader field study to trace the distribution of the ^{14}C bomb peak in many specimens across the North Sea.

[1]   C. Welte et al., Anal. Chem. 88 (2016) 8570
[2]   R. Witbaard, NETH J SEA RES 33 (1994) 91

[1] *NIOZ Royal Netherlands Inst. for Sea Research*
[2] *Inorganic Chemistry, ETHZ*

AN IMPROVED GLOBAL CARBON BOX MODEL

Inferring ^{14}C production changes from tree ring data

S. Arnold, L. Wacker, M. Christl

Changes in the atmospheric concentration of ^{14}C are recorded in tree rings and can be measured and verified over the past 10000 years. Recent studies show that unexpected peaks of ^{14}C occurred at certain times in the past, e.g. in 775 AD, and can also be found in trees [1]. For a better understanding of these peaks and to estimate the additional ^{14}C production needed to explain these events, we redesigned an existing carbon box model [2].

An existing box model [2] was improved by separating the Northern- and Southern Hemisphere to simulate the interhemispheric offset seen in tree ring data. The different distribution of land and ocean, the varying growth seasonality and the interexchange between the hemispheres in the atmosphere and ocean were considered. The evolved box model (Fig. 1) allows simulating the expected ^{14}C concentration in the different boxes on both Hemispheres.

In order to investigate ^{14}C production events and to compare them with tree ring data, we implemented an iterative X^2-test routine which fits the model to measured data. The routine first adapts the vertical and horizontal alignment (11-year solar cycle), then magnitude, date and duration of the ^{14}C event are optimized (Fig. 2). After several iterations the best fit model output is displayed together with the tree ring data.

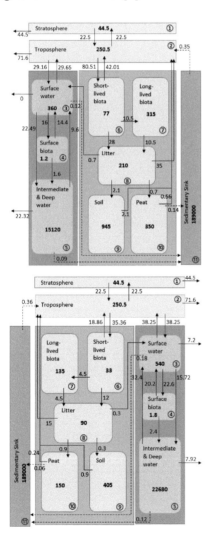

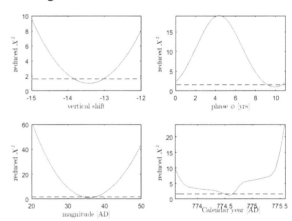

Fig. 2: Plots of X^2-tests from the fitting procedure to a possible ^{14}C peak event. The vertical and horizontal alignment, the magnitude and date of the possible event are evaluated iteratively.

Fig. 1: *Carbon box model with Northern- (top) and Southern Hemisphere (bottom) separated.*

[1] F. Miyake et al., Nature 486 (2012) 240

[2] D. Guettler et al., EPSL 441 (2015) 290

BIOMEDICAL APPLICATIONS OF ^{14}C

Validation of AMS technology for pharmacokinetic studies

D. De Maria, S. Fahrni[1], F. Lozach[2], C. Welte, L. Wacker, H.-A. Synal

Pharmacokinetic studies play an important role in the development process of new pharmaceutical compounds, where relevant metabolites have to be identified in so-called ADME (Absorption, Distribution, Metabolism, and Excretion) studies. Usually, these studies consist in the administration of a ^{14}C-radiolabeled drug followed by the analysis of biological samples by means of liquid or microplate scintillation counting. The exceptional sensitivity of AMS for the detection of rare isotopes fractions allows to significantly reduce the administrated amount of radioactive material in a microtracer approach. The applicability of AMS technologies in this field was investigated in collaboration with Novartis (Basel, Switzerland) by comparing the results from a conventional ADME study directly to those obtained by AMS-gas analysis.

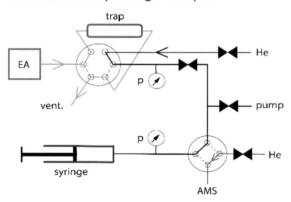

Fig. 1: *Schema of the setup used for routine gas measurements with the MICADAS at LIP.*

Samples of different biomatrices, consisting of urine, feces and plasma, were prepared using ultra high performance liquid chromatography (UHPLC) and measured at the MICADAS facility with the setup for routine gas measurements (Fig. 1). An Elemental Analyzer (EA) with a sample-feeding wheel for 80 samples was used as a combustion unit for the online analysis of the biological samples. AMS standards (Oxalic acid 2) and background (phthalic anhydride) samples were measured regularly for quality control and to ensure stable measurement conditions. Since a precise mass determination is required for a proper evaluation of the sample's activity, acetanilide was used for carbon mass calibration of the EA.

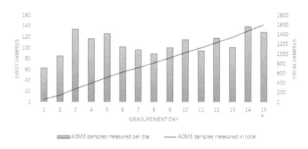

Fig. 2: *Overview of the number of UHPLC samples measured per day. In total 1608 biological samples were measured over 15 days.*

AMS measurements were performed over a period of three weeks, for a total of 1755 samples (1608 samples, 74 background/acetanilide and 73 standard samples) measured within 15 consecutive workdays (Fig. 2) proving the exceptional sample throughput and measurement stability reachable with EA-AMS.

Results obtained by ADME studies compared with those from AMS measurements reveal that both chromatograms are in agreement regardless of the fact that by the AMS-microtracer approach, a 40 000-fold lower radioactivity was dosed. However, in order to make this technology more competitive, significant technical improvements are required, starting from an increased samples throughput and more automatized processes.

[1] *Ionplus AG, Dietikon*
[2] *Novartis, Basel*

MALEVICH (?) PAINTING ANALYSED

Presence of 'Bomb peak' ^{14}C confirms suspicions of art researchers

I. Hajdas, G. Heydenreich[1], D. Blumenroth[1], S. Dietz[1]

Kasimir Malevich (1878 - 1935) revolutionized art by painting monochrome geometric figures. He founded the radical Suprematist movement. The 'Black Square' painted in 1915 (Tretyakov Gallery, Moscow) is its most famous representation. This symbolic motif has been repeated by the artist in his later works. For nearly four decades 'Black Rectangle, Red Square' (Fig. 1), which was on a display as a loan to the Wilhelm-Hack-Museum in Ludwigshafen, was believed to be one of the variations painted by Malevich. Wilhelm Hack was a German businessman who invested his wealth in a collection of fine art, mostly modernist paintings. The 'Black Rectangle, Red Square' was undocumented until mid-1970s, i.e. after Hack acquired it from an unknown source in the former Soviet Union. A change of location, after heirs of Hack donated this painting to the Kunstsammlung Nordrhein-Westfahlenin Düsseldorf triggered an investigation into its authenticity.

Fig. 1: *'Black square, Red Square'. Photo: Achim Kukulies.*

Art technological examination of the painting at TH Köln indicated a forgery: The surface appeared to be artificially patinated (aged) and the binding medium differed from other Malevich paintings. Moreover, X-ray analysis showed a different structure of paint layers as compared to 'Painterly Realism of a Boy with a Knapsack-Color Masses in the Fourth Dimension', which is a similar work by Malevich displayed in the Museum of Modern Art in New York. Finally, a piece of canvas was submitted for radiocarbon analysis. The measured ^{14}C content (F^{14}C=1.431±0.004) is higher than 1, which is only observed in the atmosphere after the nuclear tests of 1950/60s. Nuclear test created artificially increased levels of ^{14}C in the atmosphere that were reflected in vegetation growing at that time. Clearly, the measured F^{14}C of the canvas indicates post 1950's growing season for the fibres used to produce the painting support.

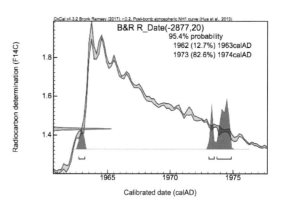

Fig. 2: *The ^{14}C content measured in the fiber of canvas corresponds to the levels observed after the onset of the 'bomb peak'.*

When compared to the atmospheric data collected since 1950s the time of canvas fiber production can be set at about 1973-74 (Fig. 2). This corroborates the documented history of this forgery.

[1] *Cologne Institute of Conservation Sciences, TH Köln, Germany*

PROBLEMS INVOLVED WITH LEAD WHITE IN PAINT

Monitoring of carbonate removal prior to dating of the organic binder

L. Hendriks, I. Hajdas, M. Küffner[1], N. Scherrer[2], S. Zumbühl[2], E. Ferreira[3], H.-A. Synal, D. Günther[4]

The possibility of dating the organic binder by ^{14}C was demonstrated in an earlier study [1]. Indeed dating the oil used for mixing of the pigments allows to make sense of canvas ^{14}C ages, which may be subject to discussion regarding the possible re-use of older canvases. The age obtained from the oil is most likely representative of the time of creation and hence is less questionable than the support material. The only pre-requisite for such an analysis is finding a suitable sampling zone, where no carbon source other than the oil is present. Hence, ideal candidates are paint locations bearing inorganic pigments. Lead white, titanium and zinc white are commonly used as white pigments and are all, by definition, inorganic and thus paint samples of choice.

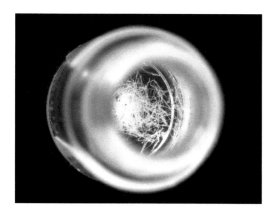

Fig. 1: *Lead chloride crystals formed after acid treatment of lead white.*

However, exceptions are carbonates, which are present in lead white. This additional carbon source is problematic for ^{14}C dating, as the provenance of the carbon is unknown and as such may compete with dating of the binder. This issue was resolved by pre-treating the sample with acid in excess, hereby forming a salt (see figure 1), water and carbon dioxide. Using a spectro Arcos ICP-OES instrument

(SPECTRO, Kleeve, Germany), the salt composition was identified as Pb and Cl with a 1:2 ratio. This result indicates that the basic lead carbonate does indeed react to form $PbCl_2$ after treatment with hydrochloric acid and hence no longer affects the dating of the binder.

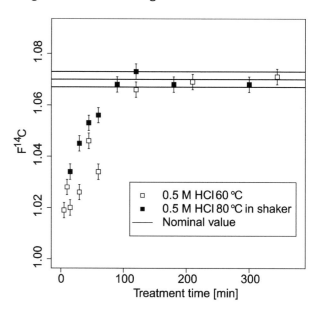

Fig. 2: *Carbonate removal efficiency in paint samples with respect to the treatment time by monitoring the obtained fraction modern.*

The removal of carbonates and formation of lead chloride was further monitored by ^{14}C dating. Figure 2 illustrates that the sample must be treated for a minimum of 2 hours to ensure complete carbonate removal.

[1]   L. Hendriks et al. Radiocarbon 60 (201) 207

[1] *SIK-ISEA, Swiss Institute for Art Research, Zurich*
[2] *HKB, Bern University of Arts*
[3] *TH Köln, University of Applied Sciences, Germany*
[4] *Laboratory of Inorganic Chemistry, ETHZ*

SAMPLE SUITABILITY USING FTIR SPECTROSCOPY

Evaluation of canvas textile prior ^{14}C analysis

L. Hendriks, I. Hajdas, M. Maurer, M.B. Röttig

In most cases when dating a painting a sample of the canvas fibers is collected. To assess sample suitability for ^{14}C analysis, it is worthwhile checking the fibers' origin. Indeed, in the 20th century, synthetic fibers were used in the production of canvases. These artificial fibers are produced from polymers that are derived from petroleum based chemistry. When such products are radiocarbon dated, they appear thousands of years old, since the ^{14}C atoms therein have decayed long ago. Thus, performing radiocarbon measurement on synthetic materials is wholly meaningless.

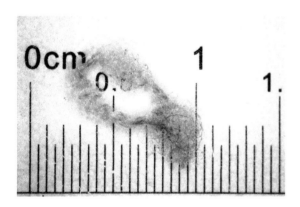

Fig. 1: *Cleaned textile fiber after soxhlet and ABA treatment.*

A good test is to collect a Fourier Transform Infrared Spectroscopy (FTIR) spectra of the material in question with the aim of identifying the presence of any synthetic materials and consequently assess whether the canvas sample is suitable for radiocarbon measurements or not.

FTIR is a highly sensitive technique in terms of identifying organic compounds. Indeed, each material has a defined chemical fingerprint in the infrared region, with characteristic IR absorption bands. For a large sample, the analysis in Attenuated Total Reflection (ATR) has the advantage of requiring no sample preparation. The sample is simply placed on the crystal and the data is collected.

Fig. 1 shows a typical cleaned canvas fiber ready for ^{14}C analysis. After measurement, it was observed that the textile was composed of synthetic material, as ages older than thousands of years were gained. The remaining material was analyzed by FTIR and the respective collected spectra are displayed in Fig. 2. The blue spectrum displays the typical absorption bands of a cellulose-based material, which is a natural fiber. In red the strong narrow bands are characteristic of the synthetic material and in this particular case polyester, also known as PET. Thus FTIR analysis confirmed the hypothesis and identified the canvas as being made from a mixture of natural and synthetic fibers, in particular, PET.

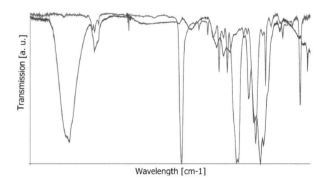

Fig. 2: *FTIR spectra of the cleaned fibers. Two distinct spectra are identified, blue is natural fibers while red is a synthetic fiber known as polyester.*

Therefore each time a canvas sample is examined for dating, a good practice is first to check whether the origin of the fibers is natural or not by collecting a FTIR spectra of the material in question and assessing its suitability for ^{14}C analysis.

GLOBAL IMPRINT OF THE 993 AD EVENT

Characterization of the 993 AD event with multiple tree-ring records

L. Wacker, S. Arnold, M. Christl, D. Galvan[1], D. Nievergelt[1], U. Büntgen[1,2]

After the discovery of the meanwhile famous 774 AD event [1], the 993 AD event [2] was the second rapid change in past atmospheric ^{14}C concentration that was found in tree-ring records. The amplitude of this event is, however, weaker and thus it is harder to detect and an independent confirmation is yet missing.

To confirm the existence and the global character of the event tree-ring samples from 990 and 1000 AD were analyzed in a total of 9 independent dendrochronologies; 6 from the northern hemisphere (NH) and 3 from the southern (SH) (Fig. 1). In 8 records a significant ^{14}C increase was found from 992 AD to 993 AD, while another record from literature [1] showed the strongest increase one year later. One NH record shows a very smooth rise from 993 to 997 AD. We think this tree ring signal might be influenced by plant physiological anomalies, i.e. the tree might have used stored carbon resources for tree-ring formation.

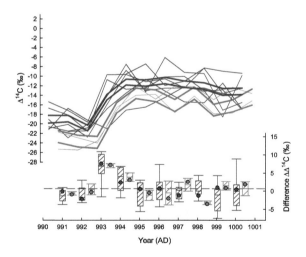

Fig. 1: *The $\Delta^{14}C$ for 6 NH records (blue) and 2 SH records is given on the top, while the year-to-year difference is given on the bottom.*

To characterize the event, a carbon cycle box model [3] fed with a short ^{14}C production burst

was fitted to our data (Fig. 2). The best fit places the event in spring (May) 993 AD with an additional production of $(8.7\pm1.4)\times10^{25}$ atoms, representing the 2-fold mean annual production.

Apparently, atmospheric $\Delta^{14}C$ in the NH record in 992 AD is lower compared to previous years. The same we observed for the 774 AD event [4]. One may hypothesize that a high proton flux was first shielding the earth from cosmic ray (thus lower production), before later higher energetic protons were then capable of producing ^{14}C.

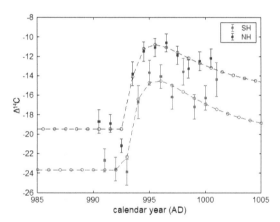

Fig. 2: *The modelled ^{14}C concentrations (open circles) are in good agreement with the measured data (solid squares) for both, the SH (red) and the NH (blue).*

The global character of the ^{14}C event was confirmed, but it was found one year earlier. We get half the amplitude compared to the 774 AD event, which is in agreement with [1].

[1] F. Miyake et al., Nature 486 (2012) 240

[2] F. Miyake et al., Nat Comm. 4 (2013) 1748

[3] S. Arnold et al., LIP Annual Report (2017) 31

[4] L. Wacker et al., LIP Annual Report (2017) 37

[1] *WSL, Birmensdorf*

[2] *Geography, Cambridge University, UK*

GLOBAL SIGNATURE OF THE 774 AD EVENT

Measurement of the radiocarbon spike in 35 dendrochronologies

L. Wacker, S. Arnold, M. Christl, D. Galvan[1], D. Nievergelt[1], U. Büntgen[1,2]

Tree-ring records store the atmospheric ^{14}C concentration in annual resolution over several hundreds to thousands of years. In a scientific collaboration with WSL (Swiss Federal Institute for Forest, Snow and Landscape Research), we measured the ^{14}C concentration between 770 and 780 AD in 35 independent tree-ring records of known calendar age. With the expected sudden increase in ^{14}C concentration between 774 and 775 AD [1] we can test the chronologies and study the global signature of the event.

In Fig. 1 the Δ^{14}C of the investigated 29 northern hemisphere (NH) and 6 southern hemisphere (SH) records are given. The records are over all in very good agreement, while we see the expected SH offset of about 4 ‰ lower values.

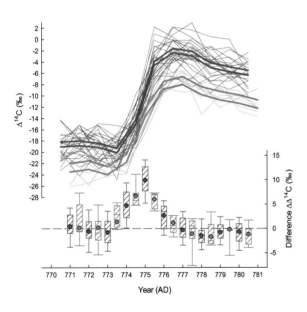

Fig. 1: Δ^{14}C for 29 NH records in blue and 6 SH records in red (top); year-to-year (bottom).

A closer look at the NH data reveals a latitude dependency (Tab. 1). The modelled additional production (Fig. 2) is larger for the northern than for the southern records on the NH. The cause of this is yet unknown. Such a latitudinal dependency is also seen when looking at 11 yr averages. While we observe the expected SH-NH offset of (-4.0±0.4) ‰, the relatively strong offset of the records from >60°N of (1.6±0.6) ‰ relative to the NH mean is rather surprising. The smaller offset for the NH records between 30° and 40°N might be explained by differences in growing season.

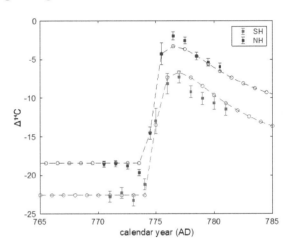

Fig. 2: Modelled ^{14}C values (open circles) agree well with measured data (solid squares) for both, the SH (red) and the NH average (blue).

	production atoms (10^{25})		Offset $\Delta\Delta^{14}$C (‰)		
NH mean	103	± 4	n.a.		
NH 60-75°N	101	± 9	1.6	±	0.6
NH 40-60°N	108	± 7	-0.2	±	0.5
NH 30-40°N	93	± 8	-1.2	±	0.6
SH 45-30°S	88	± 6	-4.0	±	0.4

Tab. 1: Modelled additional production for all records (mean) and for individual latitude zones (1st column) and zonal averages (770 - 780 AD) minus NH mean ($\Delta\Delta^{14}$C, 2nd column).

[1]   F. Miyake et al., Nature 486 (2012) 240

[1] *WSL, Brimensdorf*
[2] *Geography, Cambridge University, UK*

ROLE OF STORED CARBON IN TREE-RING FORMATION

Bomb-spike dating of early and late wood cells

J. Bitterli, L. Wacker, K. Treydte[1], D. Nievergelt[1], U. Büntgen[1,2]

Tree rings are preferably used to reconstruct atmospheric ^{14}C concentrations. However, little is known about the fixation of atmospheric CO_2 in the actual tree rings themselves [1]. Although it is known, that stored non-structural carbon (NSC) can be several years old [2], the temporal dynamics of NSC use for cellulose formation in tree stems is still unclear.

Here, we measured the ^{14}C content in cellulose extracted from two earlywood samples (EW1 and EW2) and one latewood sample (LW) (Fig. 1) for each year between 1962 and 1965. The ^{14}C concentrations for the EW1, EW2 and LW of a larch and a spruce from the Lötschental (Switzerland) were compared with atmospheric ^{14}C concentrations [3] at the time of growth (Fig. 2).

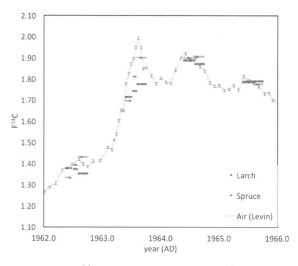

Fig. 2: *The ^{14}C concentrations (as $F^{14}C$ of EW1, EW2 and LW for a larch (green) and a spruce (orange) are compared with atmospheric concentrations.*

Fig. 1: *A scalpel was used to divide the individual tree rings into two early-wood and one late-wood fraction.*

The larch sample showed clearly lower values in the EW1 for 1963 and 1964, while the EW2 and LW values are comparable to the atmospheric values. In contrast, the cellulose of the spruce sample seems to be moderately but evenly lower throughout the growing season. We estimated that 20±5% of the larch EW1 cellulose originates from stored NSC from previous years (we took a 5-year average), while the spruce uses about 10±5% stored NSC throughout the year.

This means that only a minor portion of the cellulose originates from stored NSC, while most fixed carbon stems from atmospheric CO_2. This observation demonstrates why trees are very good archives for past atmospheric CO_2 concentrations.

However, it has to be noted, that this is a preliminary study and it seems that not all trees behave in a similar way. More measurements are required to gain deeper insight into the temporal dynamics of NSC use during cambial activity and how it may vary depending on the species and/or the environmental conditions.

[1] A. Gessler & K. Treydte, New Phytol 209 (2016) 1338
[2] A. Richardson et al., New Phytol 197 (2013) 850
[3] I. Levin et al., Tellus 62 (2010) 26

[1] WSL, Brimensdorf
[2] Geography, Cambridge University, UK

RECONSTRUCTING THE PAST WITH TREES

Past atmospheric ^{14}C concentrations between 12'325 - 11'875 cal BP

A. Sookdeo, L. Wacker, M. Friedrich[1,2], F. Reinig[3], B. Kromer[2], U. Büntgen[3], M. Paulyl[4], G.Helle[4], H.-A. Synal

Reconstructing past atmospheric ^{14}C activity is the basis for accurate ^{14}C calibrated age ranges for which tree-ring chronologies are the touchstone [1]. We have analyzed Swiss & German trees recently dendrochronologically linked to the absolutely dated Preboreal Pine Chronology (PPC). A total of 530 ^{14}C dates were measured for the 450 year period between 12'325-11'875 cal BP which contained only 12 decadal measurements before that (Fig. 1).

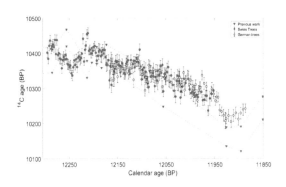

Fig. 1: *^{14}C dates for Swiss and German trees along previous work.*

Due to the lack of temporal resolution between 12'325-11'875 cal BP only 12% of the ^{14}C dates from the floating Southern Hemisphere Towai chronology (TC) were used for curve-matching. The placement of TC led to the repositioning of floating chronologies during the Bølling/Allerød/ Younger Dryas [2]. With our new ^{14}C data, which is unparalleled in temporal resolution, we can more precisely position TC.

Running a χ^2 model of the new data against TC, the position for TC shifts +10±6 cal yr with an IH offset of (35±8) ^{14}C yr (Fig. 2). Visually, this fit agrees as the ^{14}C wiggle around 12,100 cal BP aligns in both chronologies (Fig. 3).

With these results, we have successfully reconstructed past atmospheric ^{14}C activities for the end of the absolute tree-ring based part of

the calibration curve; significantly improving the precision of ^{14}C calibration throughout Bølling/Allerød/Younger Dryas.

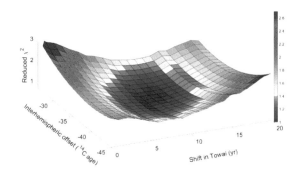

Fig. 2: *χ^2 results from varying IH offset and shifting Towai. Values greater than 1.4 (yellow and red) are outside the 95% confidence interval.*

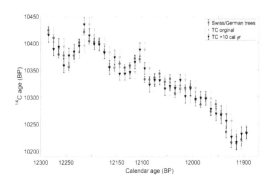

Fig. 3: *TC with its original and shifted position with Swiss and German trees. 35 ^{14}C yrs is subtracted from TC.*

[1] P. Reimer et al., Radiocarbon 55 (2013) 1923
[2] A. Hogg et al., Radiocarbon 58 (2016) 709

[1] Environmental Physics, Heidelberg University, Germany
[2] Botany, Hohenheim University, Stuttgart, Germany
[3] Swiss Federal Research Institute, WSL, Birmensdorf
[4] Helmholtz Centre Potsdam, GFZ, Germany

HIGHEST PRECISON ^{14}C FOR LATE GLACIAL WOOD

Subfossil trees ^{14}C dated with less than 3‰ error

A. Sookdeo, L. Wacker, M. Friedrich[1,2], F. Reinig[3], B. Kromer[2], U. Büntgen[3], M. Pauly[4], G. Helle[4], H.-A. Synal

Radiocarbon dating subfossil trees is of great scientific importance as it allows for reconstruction of past atmospheric ^{14}C levels, accurate calibrated age ranges and studying changing solar activity. As radiocarbon samples get older and older measurement precision becomes an issue due to reduced counting statistics and the higher impact of the blank corrections.

At ETH Zurich and Curt-Engelhorn-Zentrum (CEZ) maximum precision ^{14}C measurements were performed on 455 and 75 tree-ring samples between 12300 - 11900 cal BP, respectively. The standards showed small variability <3‰, which is in agreement with the measurement criteria. All samples reached 200000 counts.

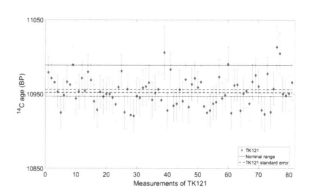

***Fig. 2**: ^{14}C age for 81 reference samples (TK121).*

Comparison of overlapping samples measured at both ETH Zürich and CEZ show no discrepancies with a reduced χ^2_{red} of 1.06 (n =72; Fig. 3). Sample errors reported by both labs are less than 3‰.

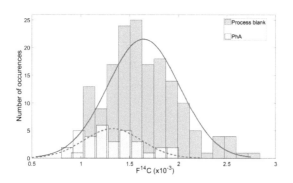

***Fig. 1:** Distribution for 164 process blanks and 33 PhA blanks.*

Chemical ^{14}C blanks (PhA) compared to process wood blanks showed no statistical difference (p >> 0.05; Fig. 1). The mean F^{14}C for the process blanks is 0.0016±0.00003 (51.7k ^{14}C yrs), and for PhAs is 0.0013±0.00005 (53.3k ^{14}C yrs), respectively.

Reference material for this time period was measured 81 times with an average ^{14}C age of (10954±2) cal BP, within one sigma of values reported by other labs [1] (Fig. 2).

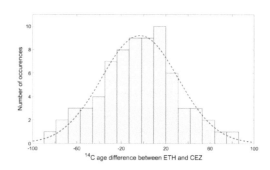

***Fig. 3:** Distribution of difference between ETH Zürich and CEZ.*

To date these are the highest precision ^{14}C measurements recorded on late glacial trees confirmed by good reproducibility between labs.

[1]   A. Hogg et al. Radiocarbon 55 (2013) 2035

[1] *Environmental Physics, Heidelberg University, Germany*
[2] *Botany, Hohenheim University, Stuttgart, Germany*
[3] *Swiss Federal Research Institute, WSL, Birmensdorf*
[4] *Helmholtz Centre Potsdam, GFZ, Germany*

CALIBRATION OF THE OXALIC ACID 2 STANDARD

Re-evaluation of the New Oxalic Acid Standard SRM-4990C with AMS

L. Wacker, S. Bollhalder, A. Sookdeo, H.-A. Synal

The New Oxalic Acid standard (NOX) was introduced in 1980 to replace the primarily Oxalic Acid (OX) standard. The NOX standard was calibrated against the OX by means of decay counting measurements. However today, most radiocarbon measurements are performed with AMS, a method that is more powerful as significantly less material is required and the measurement time is reduced. Here, we give an update on the calibration of the NOX against OX performed by AMS.

Over 300 radiocarbon measurements of OX and NOX were used to compile a precise and accurate calibration that outperforms the original calibration in precision. The mean NOX/OX ratios measured per magazine (mean of 3-4 NOX and 3-4 OX) are given in Figure 1.

The scatter of the 46 ETH measurements is 0.94‰. The high quality of the measurements is reflected in a X^2_{red} of 0.63 (n=46) that indicates we most likely over-estimate the errors. Likely, the estimated external errors of 1.0‰ to 1.2‰ for the highest-precision measurements are somewhat too large.

The mean value for the $\delta^{13}C$ (PDB) is 1.53 ±0.10‰ (Fig. 2) with a scatter of 0.7‰. This is in reasonable agreement with the consensus value of 1.49±0.05‰ determined in 1983 [1] and reflects the high stability of the MICADAS measurements.

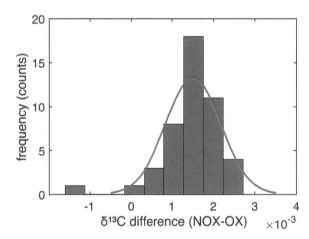

Fig. 2: *The distribution of $\delta^{13}C$ are given for the 46 highest precision measurements.*

The mean NOX(-25)/OX(-19) calibration performed at ETH Zurich is with 1.2732±0.0002 in good agreement with the accepted nominal ratio of 1.2736±0.0004 [1], but twice as precise.

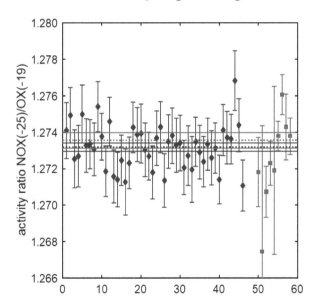

Fig. 1: *The 46 new calibration measurements performed at ETH are given in blue and the previously existing nine calibration measurements by different laboratories in red. The mean values are in agreement (horizontal lines).*

[1] W.B. Mann, Radiocarbon 25 (1983) 519

MICROSCALE RADIOCARBON ANALYSIS USING EA-AMS

Analysis of small and ultra-small combustible organic ^{14}C-samples

C. Welte[1], L. Hendriks, L. Wacker, N. Haghipour[1], T.I. Eglinton[1], H.-A. Synal

The direct coupling of an elemental analyzer (EA) to the gas ion source of the MICADAS allows the analysis of small (<100 µg) and ultra-small (<10 µg) radiocarbon samples. The increasing demand for such microscale samples leads to the need for an optimized protocol for sample preparation as well as a suitable data correction strategy. As samples are wrapped in vessels for combustion in the EA, a constant contamination characterized by a mass (m_c) and a F^{14}C is mixed with the sample. With decreasing sample size the influence of this contamination becomes increasingly significant and requires correction. The model of constant contamination allows to assess both, m_c and F^{14}C [1], but involves the labor-intensive preparation of two sets of standards for each parameter. Using the EA alone allows to determine m_c, which is decisive for minimizing the contamination.

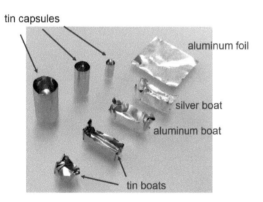

Fig. 1: *Overview of vessel shapes and materials.*

Different vessel materials (aluminum, silver, tin) as well as commercial aluminum foil (see Fig. 1) were combusted in the EA in order to assess their carbon content. Furthermore, different cleaning procedures were compared: vessels were cleaned in DCM, acetone and fully deionized water using an ultrasonic bath.

Aluminum and tin boats were combusted at 500°C. Aluminum boats were cleaned using a vacuum-oven. For comparison, uncleaned vessels were also analyzed.

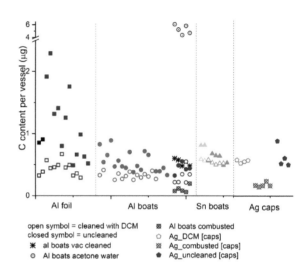

Fig. 2: *Comparison of C contamination mass introduced by different vessel materials and cleaning procedures.*

The results are depicted in Fig. 2. Dashed vertical lines indicate different vessel types. Different colours within sections represent different measurements days (i.e. replicates). For Al boats, Al foil and tin boats DCM cleaning decreases m_c. Pre-combustion leads to lower m_c for Al boats and Ag capsules. The lowest contamination is found for combusted Al boats with a mean of 0.10±0.02 µg C per vessel. For silver best results were also achieved when pre-combusting the vessels (0.17±0.02 µg C). Tin boats generally show a higher contamination (0.58±0.04 µg C).

[1] U. Hanke et al., Radiocarbon 59 (2017) 1103

[1] Geology, ETHZ

ONLINE COMPOUND SPECIFIC RADIOCARBON ANALYSIS

Analytical challenges and applications

N. Haghipour[1], B. Ausin[1], M. Usman[1], N. Ishikawa[1], C. Welte[1], L. Wacker, T.I. Eglinton[1]

With the recent developments in Accelerator Mass Spectrometry (AMS) online radiocarbon (^{14}C) gas measurements, it is possible to analyze samples as small as 5 µg C [1]. With decreasing sample size, it is crucial to test the reproducibility of data and proper blank assessment for constant contamination correction [2].

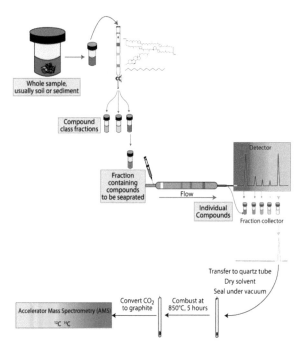

Fig. 1: *Scheme showing multiple steps involved in purification of different classes of compounds for radiocarbon analysis (modified after Ingalls et al, 2005).*

Low recovery accompanied by multiple steps during purification procedures are the most demanding analytical challenges for Compound Specific Radiocarbon Analysis (CSRA, Fig1). So far, the majority of compound specific analyses in earth sciences have been measured offline using a vacuum line for tube combustion and subsequent ^{14}C measurement with a cracker system coupled to an AMS. Compared to online measurements this is a time-consuming and expensive method.

In this study, we intend documenting processing blanks from different classes of compounds such as *n*- alkanes, amino acids, alkenones and *n*-fatty acids using an Elemental Analyzer (EA) coupled with AMS at ETH-Zurich. Furthermore, we investigate how reproducible and robust the results are when dealing with compounds smaller than 10 µg and low in concentration (> 15 ka).

The data from online measurements show that ^{14}C levels of processed and non-processed blanks are very similar to the values obtained from offline measurements making EA-AMS a promising alternative to offline measurement for CSRA. However, the magnitude of uncertainty has to be carefully considered in order to provide meaningful information and avoid misinterpretation of data.

Regardless of the specific application, in the case of small compound sample sizes (< 10 µg C) with low concentration duplicate measurements are important, as the magnitude of uncertainty derived from error propagation during data processing can be very high. This remains a big challenge in some classes of compounds.

[1] M. Ruff et al., Radiocarbon 52 (2010)
[2] U. Hanke et al., Radiocarbon 51 (2017)

[1] *Geological Institute, ETHZ*

COMPOUND-SPECIFIC ^{14}C ANALYSIS IN ECOLOGY

Do trees recycle chlorophyll?

N.F. Ishikawa[1,2], R. Nyffeler[3], N. Haghipour, T.I. Eglinton[1], N. Ohkouchi[2]

There is growing evidence that terrestrial higher plants trade organic carbon among individuals via belowground networks [1]. However, it is still unknown why and how they use such soil carbon, although, there is plenty of carbon available in the atmosphere. This study tested the hypothesis that deciduous plants such as oak and beech biosynthesize complex compounds including chlorophyll *a* by recycling the soil organic matter.

Fig. 1: *A leaf specimen that is allowed to be dedicated for compound-specific radiocarbon analysis (CSRA) of chlorophylls.*

We used historical leaf specimens (collection year: 1942-2007) in the University of Zurich to date ^{14}C age of bulk leaf and chlorophyll *a*. In addition, contemporary leaf samples and soil organic matter were collected from the Laegern site in Switzerland in summer 2016. Chlorophyll

a and its derivatives were extracted from these samples in accordance with the method we previously published [2].

Radiocarbon offers a unique and useful signature as the nuclear bomb testing in the 1950-60s drastically increased ^{14}C concentration in atmospheric CO_2 and it has continuously been decreasing since then. If chlorophyll *a* is partly synthesized by recycling the soil organic matter, a large variation in ^{14}C concentration between the bulk leaf and chlorophyll *a* is expected. Our preliminary data have already supported this hypothesis and further analysis will be made in the near future in order to fully accomplish the aim of this research project.

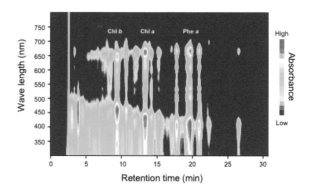

Fig. 2: *Representative HPLC chromatogram of phytosynthetic pigments extracted from plant leaf.*

[1]   R.G. Klein et al., Science 352 (2016) 342
[2]   N.F. Ishikawa et al., Biogeosci. 12 (2015) 6781

[1] *Geology, ETHZ*
[2] *Department of Biogeochemistry, Japan Agency for Marine-Earth Science and Technology*
[3] *Department of Systematic and Evolutionary Botany, University of Zurich*

RECONCILING TEMPORAL BIASES AMID PALEORECORDS

Application of specific ^{14}C-based chronostratigraphies

B. Ausín[1], C. Magill[1,2], N. Haghipour, L. Wacker, T. Eglinton[1]

Marine sedimentary sequences constitute one of the richest and most continuous archives to study past climate variability. Paleorecords based on the distribution and geochemical signature of organic carbon (OC) provide key information about productivity variations, deep basin conditions and organic matter origin. Typically, the temporal control for these proxy-records is based on radiocarbon (^{14}C) ages of co-occurring planktonic foraminifera (Fig. 1). Therefore, it is assumed that different sediment components (i.e. OC and foraminifera) that reside within the same sediment horizon were formed and deposited synchronously.

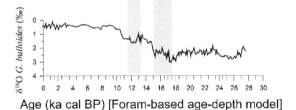

Fig. 1: *Proxy reconstruction for sediment core SHAK-06-5K. Chronostratigraphic control based on foraminifera ^{14}C ages only.*

However, we report bulk sediment OC ^{14}C ages featuring a consistent down-core offset against foraminifera ^{14}C ages of (1200±200) yr in a sediment core retrieved from the Iberian Margin (Fig. 1). These results imply the presence of allochthonous and pre-aged organic material owing to processes such as lateral transport, fluvial discharges, and resuspension of bottom sediments. Such effects have profound

implications for interpretation of multi-proxy signals (Fig. 2). Application of the OC-based chronostratigraphy to the OC-derived proxy records demonstrates that such temporal biases dramatically compromise a robust interpretation of resulting signals. Such implications might be of critical importance when assessing climate variability during abrupt climate events like the Younger Dryas (YD), the Bølling-Allerød (B/A), and the Heinrich Stadial 1 (HS1).

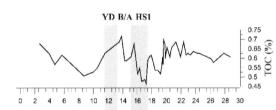

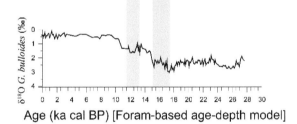

Fig. 2: *Proxy reconstruction for sediment core SHAK-06-5K. Chronostratigraphic control based on foraminifera-^{14}C ages for foraminifera-derived $\delta^{18}O$, and OC-^{14}C ages for TOC.*

Here, we demonstrate that the development and the application of specific ^{14}C chronostratigraphies provide an opportunity to reconcile temporal biases among proxy records and assess their temporal and geographical reliability.

---

[1] Geology, ETHZ
[2] Lyell Centre, Heriot-Watt University, UK

TERRESTRIAL ORGANIC MATTER EXPORT

Decoding organic matter signatures of the Godavari Basin

M.O. Usman[1], M. Lupker[1], N. Haghipour[1], L. Wacker[1], L. Giosan[2], C. Ponton[3], T. Eglinton[1]

Radiocarbon (^{14}C) measurements on individual n-alkyl lipids (e.g. fatty acids, alkanes, alcohols etc.) extracted from marine sediments have been widely used for biomarker provenance determination and as proxies for paleoclimate [1]. Long-chain ($\geq$ nC24) fatty acids found in sediments are generally derived from the epicuticular coating of vascular plant leaves [2], and incorporate the ^{14}C signature of the CO_2 from which they are synthesized. The encoded CO_2 can take several paths before being buried within continental margin sediments. While some may have been transported almost instantaneously, others may have resided in terrestrial reservoirs such as soils, flood plains etc. for several years before being eroded and transported. The Godavari Basin (Fig. 1) represents one of the former where OC transfer from source to sink is almost instantaneous owing to the basin morphology.

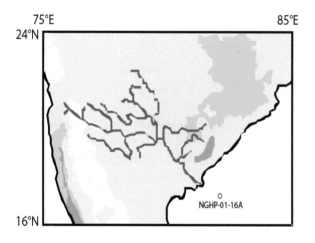

Fig. 1: *Map of Peninsular India showing the sample location (core NGHP-01-16A).*

Here, we produce a ^{14}C profile of n-alkanoic (fatty) acid from a sediment core spanning the last 17000 years. The down-core ^{14}C profile of the Godavari core shows a trend to younger ages towards the present as expected in sediments with little or no post-depositional reworking.

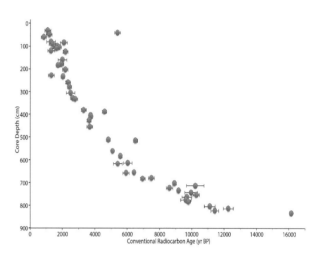

Fig. 2: *Long-chain fatty acid ^{14}C profile from the Godavari Delta core.*

Initial results highlight the potential of compound-specific radiocarbon analysis as an independent constraint on sediment chronology as well as in deciphering environmental signatures encoded in terrestrial organic matter.

[1] T. Eglinton et al., Science 277 (1997) 796

[2] G. Eglinton & R. Hamilton, Science 156 (1967) 1322

[1] *Geology, ETHZ*
[2] *WHOI, MA, USA*
[3] *Caltech, CA, USA*

CARBON STABLITY ON A HAWAIIAN CLIMATE GRADIENT

Radiocarbon dating of RPO fractions and lipid biomarkers

K. Grant [1], V. Galy[2], N. Haghipour[3], L. Wacker, T. Eglinton[3], L. Derry[1]

Soil organic carbon (SOC) is a heterogeneous mixture of carbon compounds with varying reactivity. Changing environmental conditions can lead to destabilization of SOC through disruption of protection mechanisms. We tested the impact of soil redox variability using Hawaiian andisols derived from a 400 ka Pololu (basaltic) lava flow formed on a precipitation gradient on Kohala Volcano (Hawaii, USA). Increasing precipitation leads to more frequent saturation and extensive iron loss (Fig. 1).

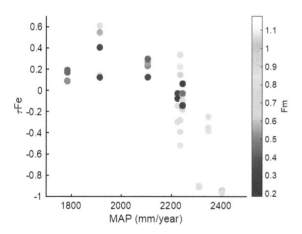

Fig. 1: *As precipitation increases across the gradient, Fe is lost from the soil ($\tau Fe < 0$). Radiocarbon as measured by Fraction Modern (Fm) is seen to generally be older in the drier region.*

Soil profiles (~1m) were sampled by genetic horizon and mineral soil horizons (50-70 cm) were analyzed on the Ramped PyRox (RPO) system at Woods Hole NOSAMS facility. Radiocarbon and stable carbon isotopes were measured on each RPO fraction and the *n*-alkanoic acids. RPO analysis gives an activation energy distribution for each thermal fraction collected and is a measure of thermochemical stability [1].

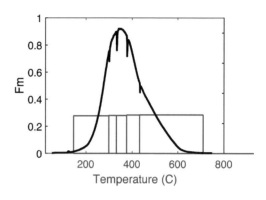

Fig. 2: *RPO thermogram showing uniform radiocarbon ages of the thermal fractions.*

RPO analyses have uniform age distributions (Fig. 2), suggesting that each thermal fraction contains a mixture of carbon compounds with the same activation energy distribution ($p_0(E)$). The short chain fatty acids (SCFA) (C_{16}-C_{18}) had Fm values ranging from 0.90 to 0.76 and long chain fatty acids (LCFA) (C_{24}-C_{32}) had Fm values of 0.63 and 0.18 for the Fe depleted and enriched sites, respectively. At all sites, average SCFA are younger than the average LCFA. However, at the Fe depleted site, the Fm values of both the SCFA and the LCFA are much higher indicating faster turnover of microbial-derived and plant-derived SOC. We suggest Fe oxides stabilize LCFA on longer time scales. Combining RPO and biomarker analysis gives a thermal, age, and structural spectrum, which provides a new perspective on SOC stability.

[1]   Hemingway et al., Biogeosciences 14 (2017) 5099

[1] *EAS, Cornell University, USA*
[2] *Marine Geochemistry, WHOI, USA*
[3] *Geology, ETH*

RIVERINE FIRE-DERIVED BLACK CARBON (BC)

Evidence for the refractory nature of river BC at global/regional scales

A I. Coppola [1], D.B. Wiedemeier[1], V. Galy[2], N. Haghipour[3] U.M. Hanke[1], G.S. Nascimento[3], M. Usman[3], T.M. Blattmann[3], M. Reisser[1], C.V. Freymond[3], M. Zhao[4], B. Voss[2], E. Schefuß[2], L. Wacker, B. Peucker-Ehrenbrink[2], S. Abiven[1], M.W.I. Schmidt[1], T.I. Eglinton[3]

Rivers annually carry about 0.028 Gt black carbon (BC) to the ocean, which is about 10% of the total dissolved organic C flux. These studies fill a key gap in our knowledge about the proportion, quality and age of fire-derived black carbon (BC) that survives river transport and export to the ocean.

Project 1: Global River Assessment Fire-derived combustion residues, i.e. BC, are the slowest cycling component of the biospheric C cycle identified up to now. Constraining the magnitude and dynamics of BC releases from the land to the aquatic environment and potential mineralization to CO_2 is critical given on-going and transformative changes in land use, hydrologic regimes, fire frequency, and severity of coastal erosion processes.

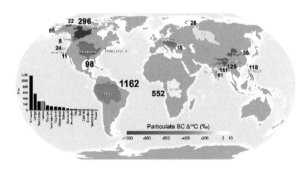

Fig. 1: *PBC fluxes (Gg yr $^{-1}$) and PBC $\Delta^{14}C$ values.*

We examined the concentration and radiocarbon (^{14}C) content (Fig. 1) of particulate BC (PBC) in 18 large and small rivers in order to assess basin-scale dynamics and estimate global flux. We show that PBC is correlated to suspended sediment yield, indicating that PBC export is primarily controlled by erosional processes. ^{14}C measurements reveal that riverine PBC is not exclusively derived from recent biomass burning, with the presence of extensively pre-aged PBC (up to 17,000±780

^{14}C yr) in several high latitude rivers. The global, river flux-weighted ^{14}C age of PBC delivered to the ocean (3,700±400 ^{14}C yr) implies extended storage in soils prior to mobilization and export.

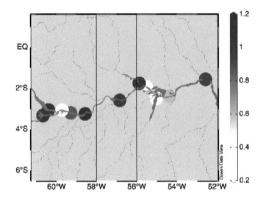

Fig. 2: *Amazon River F^{14}C values.*

Project 2: Amazon River Dissolved Organic C (DOC) in river samples from the Amazon were evaluated for $\Delta^{14}C$ to understand river dissolved BC (DBC) cycling. Despite fire-history, DBC fire-derived compounds are continuously exported [1]. This is the first study to identify the main processes behind the release and turnover of BC (Fig. 2) along the Amazon river-to-ocean aquatic continuum. DOC characterizations and $\Delta^{14}C$ DBC values were paired to determine the source and age of DBC. This ongoing project will identify the main processes behind the release and turnover of DBC using a conservative mixing model.

[1] T. Dittmar et al., Nat. Geosci. 5 (2012) 618

[1] Geography, University of Zurich
[2] Geology, ETHZ
[3] WHOI, Woods Hole, USA
[4] University of China, China

RIVERINE CARBON EXPORT IN THE CENTRAL HIMALAYA

Tracing the source of organic carbon in suspended river sediments

L. Märki[1], M. Lupker[1], N. Haghipour[1], C. Welte, T.I. Eglinton[1]

The export of organic carbon in river sediments from the Central Himalaya and its burial in the Bay of Bengal plays an important role on the global carbon cycle. In this project, we investigate the controls of topography, climate and erosion on the riverine export of organic carbon (OC). A set of river surface suspended sediment samples of five different sites in Central Nepal covering a large range of climatic conditions, and geomorphic settings is used (Fig. 1). Samples were taken during the monsoon seasons of the years 2015 and 2016 by filtering river surface water. Radiocarbon (^{14}C) dating is used to geochemically characterize the OC in suspended sediments and thus gain information about its origin.

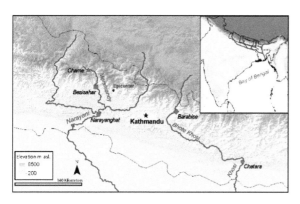

Fig. 1: *Map of Central Nepal showing the sampling sites and overview map (top right) indicating the study area in red.*

First results of ^{14}C dating show a clear difference between sediments in streams in the Upper Himalaya and in rivers of southern Nepal (Fig. 2). In the higher Himalaya the OC of suspended river sediments show relatively old ages. In contrast, the two southern rivers (Narayani and Khosi) carry more modern OC. At the sampling station in Barabise we observe an interesting evolution of the ^{14}C signature, with very young ages in June and a significant drop with the ongoing monsoon.

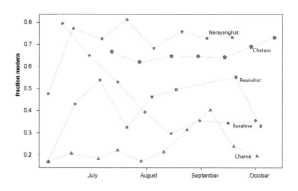

Fig. 2: *Fraction modern of suspended river sediments at different sampling sites in the monsoon seasons 2015/2016.*

This data suggests that the main sources of OC vary in different environments. The OC exported from the Upper Himalaya probably mostly originates from sedimentary rocks and from old altered soils. Rivers get loaded with younger OC when flowing through the greener lesser Himalaya, where modern soil organic matter is available. We interpret the evolution of the ^{14}C signature in the Bhote Khosi River as a consequence of the Gorkha earthquake, which hit Nepal in April 2015 and caused a large number of landslides in steep valleys east of its epicenter [1]. The Bhote Khosi valley was severely impacted by earthquake-triggered landslides. These landslides probably released a large amount of young OC from surface soils, which was exported by the river in the beginning of the monsoon, shortly after the earthquake.

[1] K. Roback et al., Geomorphology 301 (2018) 121

[1] Geology, ETHZ

RIVERINE LUZON ISLAND ORGANIC CARBON

Unexpected twists and turns in sample measurements and results

T. Blattmann[1], A. Winter[2], E. Tinacba[3], N. Haghipour[1], L. Wacker, M. Plötze[4], F. Siringan[3], T. Eglinton[1]

Southeast Asia is the world's hotspot for organic carbon land-ocean export (90 TgC/a) followed by the Amazon (31 TgC/a), the Indian subcontinent (30 TgC/a), and the Congo (13 TgC/a) [1]. Among the archipelagos of Southeast Asia, Luzon Island of the Philippines has remained one of the most unexplored areas in the tropics with regard to riverine organic carbon export. Anthropogenic activities have led to the conversion of forest landscapes to farmland leading to massive soil erosion problems and rapid changes in the soil organic carbon content [2, 3]. Here, we have investigated sedimentary organic carbon from the Luzon rivers and in the process encountered unexpected challenges on the EA-IRMS-AMS system routinely used at ETH.

The combustion behavior of Luzon sample material on the EA resulted in a strong flowrate decrease of the carrier gas. The point of blockage was found at the inlet to the IRMS and later at the inlet into the AMS. The exact reason for this abnormal blockage behavior remains unclear. The samples analyzed contain volcanic ash, inorganic phosphates, and high amounts of magnetite. Calcium phosphates might be volatilized and phosphate deposited in the capillary. An alternative explanation is the decomposition of magnetite to iron chloride during the hydrochloric acid fumigation preparation procedure, since Iron (III) chloride has a low boiling point (<400°C).

By adjusting the sample preparation for removing the carbon of carbonate and exploring the traditional sealed-tube combustion approaches, these hurdles are being overcome. A picture is emerging of the contributions of the type and age of the organic carbon being supplied to the rivers of Luzon. Surprisingly, contributions of organic carbon from agricultural soils may not be as high as hypothesized, despite pervasive anthropogenic activity in the lowlands. This may be due to seasonality or the organic carbon-depleted nature of agricultural soils.

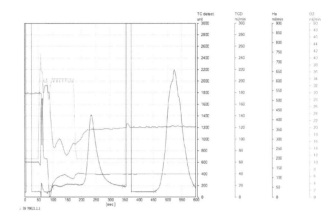

Fig. 1: *The combustion behavior of Luzon River sediments leads to a reduction in gas flow (red+blue) and a retardation of the carbon evolving from the sample (last peak in black). Additionally, the detector response from the thermal conductivity detector is altered.*

[1]   R. Schlünz & R. Schneider Int. Journ. Earth Sciences 599 (2000) 88

[2]   T. Yoneyama et al., Soil Sci. Plant Nutr. 599 (2004) 50

[3]   J. Principe, Int. Arch. Photog. Rem. Sens. Spa. Inf. Sci. 193 (2012)

[1] Geology, ETHZ

[2] Dr. Heinrich Jäckli AG, Baden, Switzerland

[3] Marine Science Institute, University of the Philippines, Quezon City, Philippines

[4] ClayLab, Institute for Geotechnical Engineering, ETHZ

TRACING DEEP SEA ORGANIC CARBON TRANSPORT

A two year ^{14}C sediment trap time series from the South China Sea

T. Blattmann[1], Z. Liu[2], B. Lin[2], X. Zhang[2], Y. Zhang[2], Y. Zhao[2], N. Haghipour[1], L. Wacker, T. Eglinton[1]

Organic carbon export and burial into the deep ocean represents the ultimate sink of carbon dioxide from and the source of oxygen to the atmosphere over geological time scales [1]. The South China Sea is one such modern deep sea sedimentary basin, which we have monitored from 2014 to 2016 to provide insight on the type and amount of carbon sequestered. Sinking ocean particles are intercepted by sediment traps suspended within the water column at different locations (see Fig. 1) and across depths ranging from 500 to 4000 m below the sea surface [2].

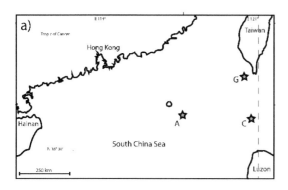

Fig. 1: *The northern South China Sea and the location of the sediment trap moorings.*

We have measured organic carbon concentration and flux as well as carbon isotopic composition (^{13}C and ^{14}C) of sinking sedimentary particles. Particularly the radiocarbon isotopic composition shows clear differences between stations and large fluctuations over time. On the one hand, organic carbon is sourced from biospheric carbon, which stands in close communication with the atmosphere. On the other hand, carbon is additionally sourced from organic carbon of petrogenic origin (i.e. eroding sedimentary rocks) from Taiwan [3]. Disentangling contributions of petrogenic carbon from carbon of immediate biospheric origin is important for budgeting carbon burial.

Reburial of petrogenic carbon represents neither a net sink of carbon nor source of oxygen to the atmosphere [4].

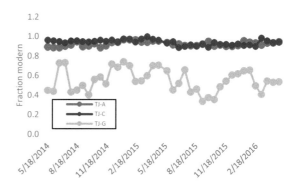

Fig. 2: *Radiocarbon isotopic composition of bulk organic carbon from the lowermost traps 50 m above the seafloor.*

Here, the influx of petrogenic carbon is controlled in large part by riverine discharge from Taiwan. However, not all petrogenic carbon pulses correlate with such terrestrial pulses and may rather be sourced from petrogenic carbon from incision of shelf sediments along submarine canyons as observed in the Mediterranean [5]. Further investigations are needed to shed more light on these processes.

[1]   R. Berner, Science 1382 (1990) 249
[2]   Y. Zhang et al., Sci. Rep. 1 (2014) 4:5937
[3]   T. Blattmann et al., LIP Annual Report (2017) 52
[4]   J. Hedges, Mar. Chem. 67 (1992) 39
[5]   T. Tesi et al., Prog. Oceanog. 185 (2010) 84

[1] *Geology, ETHZ*
[2] *State Key Laboratory of Marine Geology, Tongji University, Shanghai, China*

THE PETRIFIED TERRIGENOUS CARBON TRAIL OF TAIWAN

Anomalously depleted ^{14}C from erosion and chemical weathering

T. Blattmann[1], S-L. Wang[2] , L.-H. Chung[2], Y-P. Chang[2], N. Haghipour[1], L. Wacker, M. Lupker[1], T. Eglinton[1]

Taiwan is one of the most rapidly rising mountain fronts on Earth today. This rapid rise coupled with high erosion rates of sedimentary and metasedimentary rocks supplies abundant radiocarbon-dead kerogen to the rivers, which then emanates into the surrounding ocean [1].

The Gaoping River investigated in this study is one of the largest rivers draining Taiwan and is characterized by pronounced seasonality in precipitation and discharge with frequent typhoons. Following the powerful Typhoon Morakot in 2009, which triggered over 9000 landslides in the catchment of the Gaoping Canyon, much of the land surface covered by lush vegetation lay exposed along the steep mountain slopes [2]. Riverbed samples collected in 2010 from ten sites show strongly depleted radiocarbon signatures ranging between 5 and 30 PMC, reflecting denudation and surface exposure of shale. How much of this signature is due to the Morakot event and how much is natural background signal is subject of ongoing investigations.

Fig. 1: *Sampling along the Gaoping River.*

The dissolved inorganic carbon within the Gaoping catchment shows an anomalously low ^{14}C content ranging between 40 and 75 PMC.

These values may be explained by (a) remineralization of petrogenic carbon within the river and (b) supply of radiocarbon-dead carbon from the weathering of limestones. In the case of the former, incorporation of dead carbon along the bedrock-soil interface has been reported from Taiwan [3]. Conceivably, degradation of shale-derived organic matter in the river could also give rise to the observed DI^{14}C values. In the case of the latter, this may be induced by sulfuric acid weathering of limestone. Sulfuric acid is a degradation product of pyrite, which is a common mineral found in shales and contributes significantly to Taiwan river sulfate load [4]. Alkalinity derived from sulfuric acid weathering of limestone generates a DIC pool sourced entirely from fossil carbon. In order to disentangle these two modes of DIC generation within Taiwanese rivers, stable carbon isotope measurements will be conducted in order to pinpoint their source.

[1] R. Hilton et al., Geology 71 (2011) 39
[2] F. Tsai et al., Nat. Hazards Earth Syst. Sci. 2179 (2010) 10
[3] J. Hemingway et al., Goldschmidt Abst. (2017)
[4] A. Das et al., Geophys. Res. Lett. L12404 (2012) 39

[1] Geology, ETHZ
[2] Department of Oceanography, National Sun Yat-sen University, Kaohsiung, Taiwan

WHAT ON EARTH HAVE WE BEEN BURNING?

Reconstructing combustion history from aquatic sediments with [14]C

U.M. Hanke[1], C.M. Reddy[2], L. Wacker, N. Haghipour[3], C.M. McIntyre[3], A.I. Coppola[1], L. Xu[4], A.P. McNichol[4], S. Abiven[1], M.W.I. Schmidt[1], T.I. Eglinton[2,3]

Fire has shaped our planet Earth for about 420 million years but the utilization of fossil fuels marked the beginning of a new era, the 'Anthropocene'. Here, we studied the anoxic sediments of Pettaquamscutt River basin, RI, United States (U.S.) to investigate whether the use of different types of fuel has been recorded in terrestrial archives.

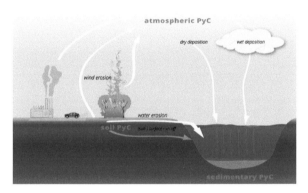

Fig. 1: *Origin of pyrogenic carbon (PyC) and its transport pathways in the environment.*

The combustion of biomass and fossil fuels is commonly incomplete and produces, in addition to carbon dioxide, solid residues known as pyrogenic carbon (PyC) or black carbon. After its formation, it can be transported by wind and water before it is either degraded or (ultimately) buried in sedimentary archives (Fig. 1). Previous research showed that there are at least two different classes of PyC, polycyclic aromatic hydrocarbons (PAHs), which are precursors of condensation and 'bulk' PyC. The latter makes up the large majority of material, yet its heterogeneous nature complicates its investigation. The benzene polycarboxylic acids (BPCAs) are operationally-defined and semi-quantitative measures of 'bulk' PyC but they have the great potential to yield robust isotopic signatures that can facilitate source apportionment to either biomass burning or fossil fuel combustion. Results of small-scale

radiocarbon measurements revealed that both combustion markers trace the consumption of fossil fuels starting in the late 1800s. Both parallel the records of fuel consumption in the U.S. until the 1970s [1]. Thereafter the records of BPCAs and PAHs tend to decrease continuously due to advent of environmental regulations and the usage of cleaner burning fuels attenuating or changing the environmental impact (Fig. 2).

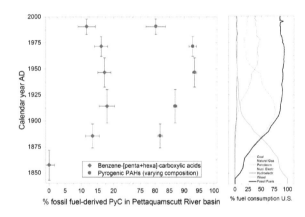

Fig. 2: *Contribution of fossil fuel-derived PyC in Pettaquamscutt River from 1850 to 2000 [1].*

Whilst PAHs are largely composed of fossil PyC (>80%), 'bulk' PyC (BPCAs) never contains more than 20%. This implies that 'bulk' PyC foremost traces local combustion practices with an additional regional source of soot from industry and transport to which PAHs are likely to be adsorbed to soot during long-range atmospheric transport.

[1]   U.M. Hanke et al., Environ. Sci. Technol. 51 (2017) 12972

[1] *Geography, University of Zurich*
[2] *Marine Chemistry and Geochemistry, WHOI*
[3] *Geology, ETHZ*
[4] *Geology and Geophysics, WHOI*

COSMOGENIC NUCLIDES

Holocene history of Triftje and Obersee glaciers

Deglaciation of northernmost Norway

Glacial stages in the Chagan Uzun Valley, Altai

The early Lateglacial around the Säntisalp

Landscape evolution of the Göschener Valley, Uri

The Glacial landscape at Wangen an der Aare

Tracking boulders with glacier modelling

Stabilisation of rock glaciers in the Tatra Mts

Chronology of terrace formation in Pratteln

First depositional age for Swiss Deckenschotter

A ~1 Ma record of glacial and fluvial deposits

Deciphering landscape evolution

Growth of relief in the Central Menderes Massif

Quantifying chemical vs. mechanical denudation

Spatial variability erosion rates in NW Iran

Soil formation in subtropical terrain

Soil erosion in a hummocky moraine landscape

Early to Late Pleistocene Death Valley fans

Slip-rate of the Ovacik fault (Turkey)

Patterns of large landslides in the Alps

^{10}Be in Black Sea sediments

^{36}Cl dating in the Nubian sandstone aquifer

The GRIP ice core ^{10}Be project

HOLOCENE HISTORY OF TRIFTJE AND OBERSEE GLACIERS

Investigated by using ^{10}Be exposure and radiocarbon dating

O. Kronig, S. Ivy-Ochs, I. Hajdas, M. Christl, C. Wirsig, C. Schlüchter [1]

Based on detailed field work combined with ^{10}Be exposure and radiocarbon dating, we reconstructed the history of glacier fluctuations of the Trifte and the Obersee glaciers for the last ca. 12 kyr [1].

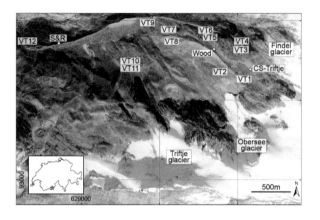

Fig. 1: *Aerial image of the study area (reproduced with the authorisation of swisstopo (JA100120)). Red dots and yellow squares show the sample locations of the ^{10}Be and the radiocarbon samples, respectively.*

The results show that a coalesced glacier completely covered the entire study area during the Egesen stadial. The first bedrock islands became successively ice-free between (12460±340) yr and (10170±420) yr (VT4, VT5, VT6, VT10, VT11). Landforms deposited during the middle Holocene were not observed in the study area.

Four ^{10}Be dates (VT2 1270±160, VT7 1590±269, VT8 1060±240, VT12 970±130 yr) and radiocarbon data of buried paleosols in the left lateral moraine of the Findel glacier (location denoted by S&R on Fig. 1; 3.2–2.2, 1.8–1.3, 1.4–0.6, 1.3–0.5 cal kyr) [2] clearly document advances of the Obersee glacier and the Findel glacier during the Göschenen II stadial.

During the Little Ice Age (LIA) re-advances, the glacier overran and/or destroyed most of the already existing depositional landforms. ^{10}Be data from the right-lateral Oberseegletscher moraine (VT1 420±170 yr, Fig. 2) and a wood sample (Fig. 1 Wood; 356-63 cal yr before 2013) from within the till testify to formation of the moraine during the LIA, as also shown on historical topographic maps available since the middle of the 19th century. Although the largest moraines in the glacier forefields are called 'Little Ice Age' moraines, they are often actually composite moraines which have been built-up during numerous late Holocene advances. Post-LIA advances were mapped based on aerial photos and available topographic maps.

Fig 2: *Sampled boulder VT1.*

[1]   O. Kronig et al., Swiss J. of Geosci. 2018 (in press)

[2]   W. Schneebeli and F. Röthlisberger, "8000 Jahre Walliser Gletschergeschichte" (1976) Verlag Schweizer Alpen Club

---

[1] Geology, University of Bern

DEGLACIATION OF NORTHERNMOST NORWAY

Surface exposure dating of major end moraines

A. Romundset[1], N. Akçar[2], O. Fredin[1], D. Tikhomirov[2], R. Reber[2], C. Vockenhuber[3], M. Christl[3], C. Schlüchter[2]

The subarctic fjord district of Finnmark county in northernmost Norway (Fig. 1) holds a well-preserved record of end moraines left by the Scandinavian Ice Sheet during the last deglaciation. The end moraines form near coast-parallel belts that can be traced over long distances.

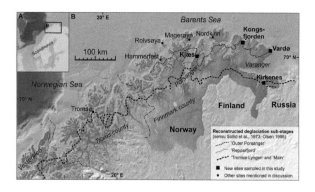

Fig. 1: *Map of the study area in northern Norway, including the new sampling locations.*

This is a palaeoglaciologically important region as it sits near the proposed border-zone between the former Scandinavian and Barents Sea Ice Sheets. However, until now the deglaciation history has few direct dates onshore. The chronology of ice front retreat has been established by correlation of ice-marginal deposits with isostatically raised shorelines (Fig. 2) and marine sediment cores.

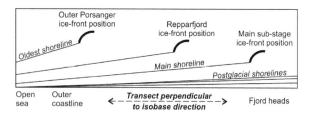

Fig. 2: *Principal shoreline diagram from Finnmark, illustrating how raised shorelines in previous studies have been mapped to major ice-marginal deposits.*

We measured the ^{10}Be (n=22) and ^{36}Cl (n=17) concentrations from boulders located at the crest of major moraine ridges at four localities: Kjæs, Kongsfjorden, Vardø and Kirkenes [1]. These are key localities for existing regional reconstructions of ice recession in this area.

Our results show that the Kongsfjorden and Vardø moraines were deposited (14.3±1.7) ka and (13.6±1.4) ka, respectively, and thus point to an Older Dryas age of the proposed 'Outer Porsanger' deglaciation sub-stage. Moraine ridges belonging to the 'Main' sub-stage near Kirkenes (Fig. 3) were dated to (11.9±1.2) ka, corresponding well to the ice retreat chronology farther west in northern Norway. Our ages suggest that the maximum Younger Dryas ice sheet extent was attained in the late Younger Dryas along a more than 500 km long stretch in northernmost Scandinavia.

Fig. 3: *Sampling the Main deglaciation sub-stage of Finnmark, shown through this study to be of late Younger Dryas age.*

[1]   A. Romundset et al., Quat. Sci. Rev. 177 (2017) 130

[1] *Geological Survey of Norway, Trondheim*
[2] *Geology, University of Bern*

GLACIAL STAGES IN THE CHAGAN UZUN VALLEY, ALTAI

Surface exposure dating of moraines with [10]Be, Altai Mountains

E. Garcia Morabito [1], C. Terrizzano[1], V.S. Zykina[2], M. Christl, R. Zech[3]

The mountains in southern Siberia (Altai, Sayan, Transbaikalia) were glaciated repeatedly during the Quaternary, and the glacial sediments and landforms there are valuable archives for paleo-environmental and paleoclimate reconstructions. However, studies focusing on glacial chronologies in these mountain ranges are still very scarce, and the extent and timing of glacier advances remains poorly constrained [1,2].

In this study we combine detailed geomorphological mapping and in situ cosmogenic [10]Be surface exposure dating of glacially-transported boulders in order to provide a paleoglaciological reconstruction of the Chagan Uzun Valley, Russian Altai Mountains. Here, extensive lobate moraine belts reflecting former glacier advances are preserved in the intermontane Chuja Basin. These former glacier advances coexisted with ice-dammed lakes and cataclysmic flood events [3].

Recently published cosmogenic ages associated with the outermost, intermediate, and inner moraines all indicate deposition ages clustered around 19 ka [4]. However, available chronological data have to be considered cautiously because of shielding effects caused by glacial lake water coverage, and given the highly dynamic geologic setting of the region.

Our data indicate that the deposition time of the outermost and intermediate moraine complexes considerably predate previously published [10]Be ages. Assuming negligible nuclide inheritance and interpreting the oldest samples from a moraine as best available estimate, a massive glaciation occurred as early as 36 ka. Subsequently, successively less extensive glacier advances occurred at 30 ka (Marine Isotope Stage 4?), 28 ka, 21 ka, and 18 ka.

This chronology indicates a last maximum extent of the Chagan Uzun Glacier during Marine Isotope Stage 3, out of phase with global ice volume records, and an onset of deglaciation around 21 ka.

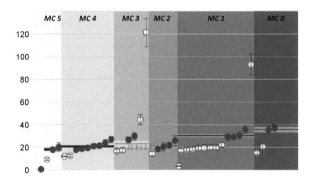

Fig. 1: [10]Be ages (calculated using Lal/Stone time-dependent scaling model) for glacial cobbles and boulders plotted for each glacier stage. The horizontal black lines and the grey bands show the mean and 1σ for each moraine complex, respectively, including ages from [2], [4], [5].

[1]   A.R. Agatova et al., STRATI (2013) 903

[2]   F. Lehmkuhl et al., Dev. Quat. Sci. 15 (2011) 967

[3]   A. Reuther et al., Geology 34 (2006) 913

[4]   N. Gribenski et al., Quat. Sci. Rev. 149 (2016) 288

[5]   A. Reuther, Relief Boden Paläoklima 21 (2007) 213

[1] Geography, University of Bern
[2] Geology of the UIGGM, Russia
[3] Geography, Friedrich-Schiller-Universität Jena, Germany

THE EARLY LATEGLACIAL AROUND THE SÄNTISALP

Glacial reconstruction and surface exposure dating using ^{36}Cl

G. Ruggia, S. Ivy-Ochs, O. Kronig, J.M. Reitner[1]

Short-term climate variations during the early Lateglacial phase are still poorly understood. due to the lack of datable landforms of this time. Most Alpine valleys were still covered by the massive valley glaciers from the retreating LGM glaciation. The Alpstein Massif (eastern Switzerland) is thought to have become an independent glacier system soon after the LGM, and was not connected and influenced by the large Rhine glacier anymore. Thus, the small glaciers in the Alpstein possibly recorded climate changes in this very early Lateglacial time. Around the Säntisalp, a detailed geomorphological and geological map was made (Fig.1) based on field observations complemented by aerial photographs, digital elevation model (DEM) and available historic and topographic maps [1].

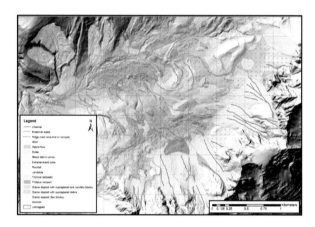

Fig. 1: Geomorphological map of the Säntisalp.

With the mapped landforms, the paleoglaciers and their ELA (Equilibrium Line Altitude) were reconstructed using GlaRe, an ArcGIS toolbox for reconstructing glaciers (Fig.2) [2]. The ELA allows to draw conclusions about paleo-temperature and -precipitation.

To determine the time of the past glacial advances, cosmogenic nuclide exposure dating of the glacial landforms will be done. During fieldwork the location of suitable boulders for cosmogenic nuclide exposure dating were mapped (Fig. 3). As the main lithology in the Alpstein area is limestone, ^{36}Cl will be the isotope used for dating.

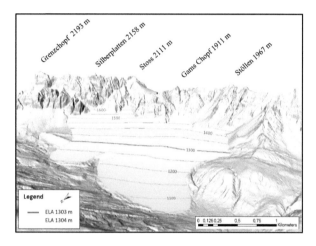

Fig. 2: Reconstructed paleoglacier and ELA elevation.

Fig. 3: Possible boulder for surface exposure dating.

[1]   G. Ruggia, Bsc Thesis, ETHZ (2017)
[2]   R. Pellitero et al., Comput and Geosci. 94 (2016) 77

[1] *Geological Survey of Austria, Vienna, Austria*

LANDSCAPE EVOLUTION OF THE GÖSCHENER VALLEY, URI

[10]Be dating questions the definition of the "Göschener" cold phases

M. Boxleitner , M. Maisch[1], M. Egli[1], D. Brandova[1], A. Musso[1], M. Christl, S. Ivy-Ochs

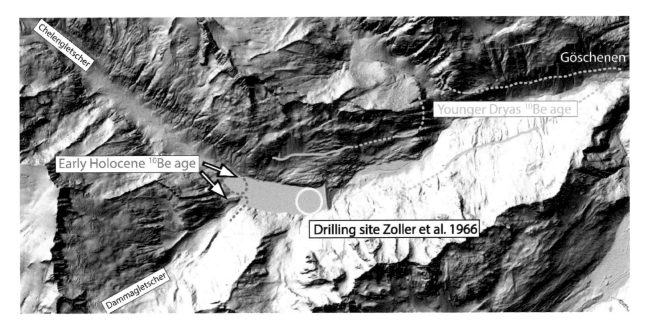

Fig. 1: *Map of the research area with relevant moraines (dotted lines: inferred moraine outlines).*

In the Göschener Valley we investigate the glacial history and landscape development and address in particular the definition of the so-called "Göschener cold phases" [1]. In a first paper we focused mainly on the evolution of soils and mires after deglaciation [2]. The disappearance of the ice as a starting point for these processes could be dated by in-situ [10]Be-dating: Prominent lateral moraines (Fig. 1) that indicate a glacier end position in today's village of Göschenen could be assigned to the Younger Dryas. We assume that the Göschener cold phase I and II (~2.5 and ~1.5 ka BP) should have left easily recognizable traits, e.g. in the pollen signal. Our results, however, could not unambiguously confirm the findings of [1] that were obtained from the analysis of pollen samples and the dating of organic remnants that were found underneath a thick scree layer during drilling explorations for the dammed lake. Zoller et al. interpreted the scree as till that was deposited during a late Holocene advance of the Chelengletscher (Göschener

phase I). Although they also considered other explanations, the Göschener cold phases since then have been commonly cited as late Holocene phases with considerable glacier advances. Because of the artificial lake the original site of [1] cannot be revisited. But ca. 1.5 km up valley moraines with boulders suitable for [10]Be dating were preserved (Fig. 1). Our results indicate that these moraines were deposited during the early Holocene and since then not overridden by another glacier advance. Consequently, Zoller's interpretation of the scree at the drilling site as till deposited by a late Holocene glacier and therefore the glaciological definition of the Göschener cold phases has to be reconsidered.

[1]   H. Zoller et al., Verhandl. Naturf. Ges. Basel, 77 (1966) 97

[2]   M. Boxleitner et al., Geomorph. 295 (2017) 306

[1] *Geography, University of Zurich*

THE GLACIAL LANDSCAPE AT WANGEN AN DER AARE

A new look at a classical glacial-geological landscape

S. Ivy-Ochs, K. Hippe, C. Schlüchter[1]

Wangen an der Aare lies in the westernmost part of the Swiss Alpine foreland (location shown on the inset of Fig. 1). The landforms and deposits in the Wangen an der Aare region and the adjacent Jura Mountains played a key role in the birth of the glacial theory in the early 19th century. The dilemma they faced was: What is the origin of the giant erratic blocks? How did they get from the High Alps to the forelands? Finally the flood theory was laid to rest and the glacial theory gained acceptance [1].

The glacial landforms at Wangen an der Aare provide a footprint of the extent of the Valais (Rhone) glacier during the Last Glacial Maximum. Direct dating with cosmogenic ^{10}Be of erratic boulders on the outermost moraine complex suggests withdrawal of the Valais glacier from the Wangen position no later than 24 ka [2]. Coarse gravels carried by meltwater streams emanating from the glacier were deposited as outwash plains to the northeast of the moraines. As the glacier withdrew, step-wise incision into the outwash fans left a flight of terraces extending down to the present flood plain of the Aare River. The terrace designated as I (Fig. 1) is the oldest. It was then incised by meltwater streams as indicated by the arrows near Oberbipp on the figure. Next, the glacier retreated a bit back from its maximum position and the dominant meltwater channel followed what is now the path of the Aare River. Terraces II and III record the stepwise retreat of the glacier back out of the Wangen region with the associated incision. The meltwater channel in the Önz River valley lost importance as well as the glacier melted back. The topographic relationships of the three main terrace levels are shown in the cross-section of the lower panel of Fig. 1.

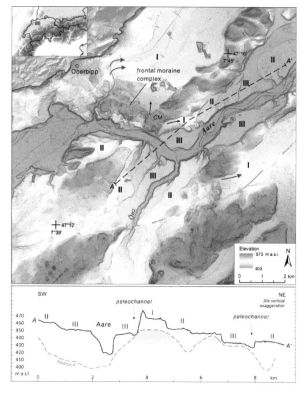

Fig. 1: *Detail of the ice marginal landforms and outwash terraces of the Wangen an der Aare region. Terrace relative ages from oldest to youngest are I, II, III. Arrows show the meltwater paths. Slope raster superimposed by a color-coded DEM (swissALTI3D) reproduced by permission of swisstopo (JA100120). Altitude is given by color code. Cross-section in lower panel.*

[1] S. Ivy-Ochs, K. Hippe, C. Schlüchter in: "Swiss Landscapes" (in press)

[2] S. Ivy-Ochs, Cuadernos de investigación geográfica 41 (2015) 295

[1] *Geology, University of Bern*

TRACKING BOULDERS WITH GLACIER MODELLING

Reconstructing LGM climate in the northwestern Alps

G. Jouvet[1], J. Seguinot[1], S. Ivy-Ochs, M. Funk[1]

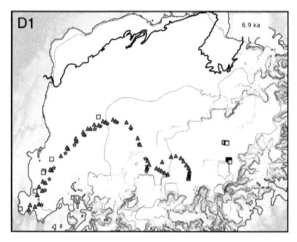

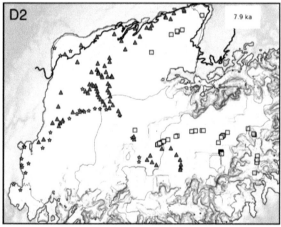

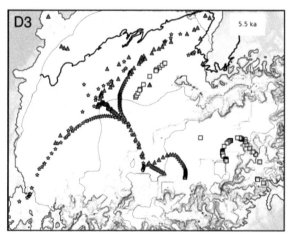

Fig. 1: *Snapshots of the modelled ice extent, and lithology markers (red Mt. Blanc granite,*

blue Arolla gneiss, green Allalin gabbro) when the volume of ice was maximum. Top: Experiment D1 'today's precipitation pattern'; Middle: Experiment D2 'corrected precipitation pattern' (southwesterly advection of moisture); Bottom: Experiment D3 'colder Jura'.

In this study, a modelling approach was used to investigate the cause of the diversion of erratic boulders from Mont Blanc and southern Valais by the Valais Glacier to the Solothurn lobe during the Last Glacial Maximum (LGM). Using the Parallel Ice Sheet Model, we simulated the ice flow field during the LGM, and analyzed the trajectories taken by erratic boulders from areas with characteristic lithologies. The main difficulty in this exercise lay with the large uncertainties affecting the paleo climate forcing required as input for the surface mass-balance model. In order to mimic the prevailing climate conditions during the LGM, we applied different temperature offsets and regional precipitation corrections to present-day climate data, and selected the parametrizations, which yielded the best match between the modelled ice extent and the geomorphologically based ice - margin reconstruction. After running a range of simulations with varying parameters, our results showed that only one parametrization allowed boulders to be diverted to the Solothurn lobe during the LGM. This precipitation pattern supports the existing theory of preferential southwesterly advection of moisture to the Alps during the LGM, but also indicates strongly enhanced precipitation over the Mont Blanc massif and enhanced cooling over the Jura Mountains.

[1]   G. Jouvet et al., J. of Glaciology 63 (2017) 487

[1] *VAW, ETH*

STABILISATION OF ROCK GLACIERS IN THE TATRA MTS

[10]Be dating of relict rock glaciers of the Lateglacial-Holocene transition

J. Zasadni[1], P. Kłapyta[2], E. Broś[1], S. Ivy-Ochs, A. Świąder[1], M. Christl

The Tatra Mountains are the northernmost, highest and coldest massif in the Carpathian Mountains. Pleistocene glaciations left an impressive landscape with a wide range of glacial and periglacial landforms. This includes rock glaciers, which are valuable diagnostic landforms of both present and past permafrost occurrence in mountain environments. In most of the previous studies rock glaciers in the Tatra Mountains were classified as relict landforms of the Latest Pleistocene age, however, several authors state that some of the high located debris bodies may contain permafrost and were active during the Holocene, e.g. during the Little Ice Age. The statement that some rock glaciers are intact is inferred form geophysical studies which suggest occurrence of discontinuous permafrost in the Tatra Mountains above 1930 m a.s.l. These contradictions about the activity status of the Tatra Mountains rock glaciers can be resolved by dating these landforms.

We apply [10]Be cosmogenic nuclide dating of rock glacier boulders in the SW sector of the High Tatra Mountains, in high elevated glacial cirques of the Kriváň mountain group (Slovakia). The investigated bodies of debris accumulation belong to the youngest (highest located) landsystem. We sampled four landforms which include one pure talus rock glacier (Fig. 1), two moraine-derived rock glaciers and one moraine type wall in the upper zone of a moraine-derived rock glacier (Fig. 2). Sampled landforms are located at 1980-2150 m a.s.l., thus above the presumed limit of discontinuous permafrost.

The mean exposure ages of the dated landforms indicates that the youngest moraines and rock glaciers in the Tatra Mountains were formed during the Younger Dryas, but final rock glacier stabilization and permafrost melting out was delayed by about 0.5 - 1.5 ka after the end of the Younger Dryas. These results testify that in the Tatra Mountains the widespread permafrost occurrence and its morphological expression in creeping of rock glacier bodies did not occur during the Holocene. Boulders found on one of the highest located rock glaciers in the Tatra Mountains have been stable for more than ten thousand years.

Fig. 1: *Sampled moraine/relict rock glacier (2130 m a.s.l.) in the Nefcerská Valley.*

Fig. 2: *Sampled relict talus rock glacier (1980 m a.s.l.) in the upper part of the Mengusovská Valley. This is a classic example of a rock glacier in the Tatra Mountains.*

[1]*Geology, AGH-UST, Kraków, Poland*
[2]*Geography, Jagiellonian University, Kraków, Poland*

CHRONOLOGY OF TERRACE FORMATION IN PRATTELN

Depth-profile dating with ^{10}Be and ^{36}Cl

A. Claude[1], N. Akçar[1], S. Ivy-Ochs, F. Schlunegger[1], P. Rentzel[2], C. Pümpin[2], D. Tikhomirov[1], P. Kubik, C. Vockenhuber, A. Dehnert[3], M. Rahn[3], C. Schlüchter[1]

The Quaternary stratigraphy of the Alpine Foreland consists of distinct terrace levels, which have been assigned to four morphostratigraphic units: Höhere (Higher) Deckenschotter, Tiefere (Lower) Decken-schotter, Hochterrasse (High Terrace) and Niederterrasse (Lower Terrace). Here, we focus on the terrace gravels at Hohle Gasse, SSE of Pratteln near Basel, which are mapped as Tiefere Deckenschotter by various authors (Fig. 1).

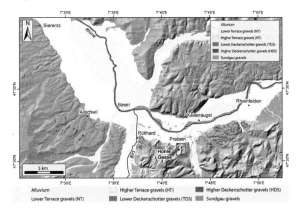

Fig. 1: Map with the Quaternary units of the Rhine River and location of the study area [1].

Sedimentological analyses indicate that gravels were transported by a braided river and deposited in a distal glaciofluvial setting. In addition, it can be shown that the majority of the clasts display multiple reworking and only a minority have a distinctly glaciofluvial shape [1].

Cosmogenic ^{10}Be depth-profile dating on seven sediment samples (Fig. 2) resulted in a minimum age of ca. 270 ka. A high blank correction made it impossible to calculate an age from the ^{36}Cl concentrations. Based on the ^{10}Be age, we suggest a classification into the morpho-stratigraphic unit Hochterrasse instead of Tiefere Deckenschotter. In addition, our results show that at Hohle Gasse, distinguishing between Tiefere Deckenschotter and Hochterrasse based on elevation is problematic. Clast morphometry and petrography could as well not help to differentiate between both units. Only absolute dating allowed a re-evaluation of the stratigraphic attribution. Our analyses further suggest that a major phase of aggradation occurred during MIS8.

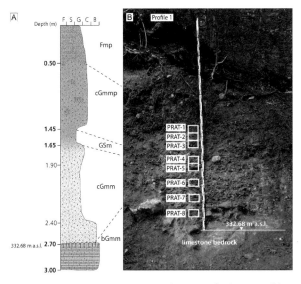

Fig. 2: A: Stratigraphic column of the profile; B: Field photograph illustrating the sampled outcrop for depth-profile dating [1].

[1] A. Claude et al., Swiss J. Geosci. 110 (2017) 793

[1] Geology, University of Bern
[2] Integrative Prehistory and Archaeological Science (IPNA), Basel University
[3] Swiss Federal Nuclear Safety Inspectorate ENSI

FIRST DEPOSITIONAL AGE FOR SWISS DECKENSCHOTTER

^{10}Be and ^{26}Al isochron burial dating quaternary terraces in Switzerland

R. Grischott, F. Kober[1], S. Ivy-Ochs, K. Hippe, M. Lupker[2], M. Christl, C. Vockenhuber, C. Maden[3]

The timing of the deposition and incision of glaciofluvial gravels that form high elevated terraces and plateaus in the northern Swiss Foreland – the so called Swiss Deckenschotter – is of special interest as they mark the onset and persistence of glacial or glacial related landscape forming processes since the Pliocene/Pleistocene boundary [1].

We applied isochron burial dating [2] for the site "Iberig" in Würenlingen representing the Lower Deckenschotter unit (Fig. 1). Samples collected from one stratigraphic horizon were analysed for their ^{26}Al and ^{10}Be content. The low inheritance of glaciofluvial sediments and subsequent long burial time result in low ^{26}Al concentrations, which makes measurement with low uncertainties a challenge [3]. Extremely low ^{27}Al content in the quartz mineral separates is an absolute necessity and requires several additional cleaning steps.

Fig. 1: *Outcrop of Lower Deckenschotter at site Iberig with marked samples.*

So far, we have processed five quartz-bearing clasts. Following criteria given in Akçar et al. [3], three samples qualified for use in the isochron-burial age model calculation (Fig. 2). We implemented two initial isochron lines, one with a slope of 8.4 (for production at depth) and the other with 6.8 (for surface production), as suggested by Akçar et al. [3]. A preliminary

estimate reveals an age range of 1.2 to 0.7 Myr, respectively, considering these two endmembers (Fig. 3). More samples are being processed to further constrain the isochron slope and the calculated age.

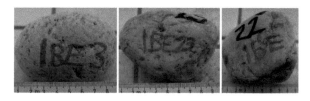

Fig. 2: *Photographs of the three samples Ibe3 (granite), Ibe23 and Ibe22 (both quartzites) used in the regression model.*

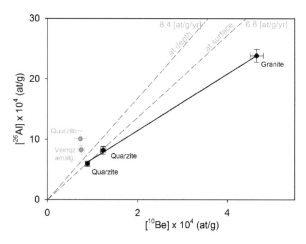

Fig. 3: *^{26}Al-^{10}Be plot with the two initial isochron lines and the dataset of five samples.*

[1]   H.R. Graf, Diss. (1993)
[2]   G. Balco and C.W. Rovey, Am. J. Sci. 308 (2008) 1083
[3]   N. Akçar et al., Earth Surf. Proc. Landf. 42 (2017) 2414

[1] *Nagra, Wettingen*
[2] *Geology, ETHZ*
[3] *Geochemistry and Petrology, ETHZ*

A ~1 MA RECORD OF GLACIAL AND FLUVIAL DEPOSITS

Surface exposure dating (^{10}Be) in Central Patagonia

E. García Morabito[1], J. Tobal [2], C. Terrizzano[1], M. Ghiglione[2] , M. Christl, J. Struck[3], L. Schweri[1], R. Zech[3]

The foreland of Patagonia provides a perfect setting to evaluate the landscape response to tectonic and climatic forcing. Valleys that were formerly occupied by glaciers often contain several Quaternary moraine and terrace assemblages that have the potential to provide paleoclimatic information extending beyond the last glacial cycle [1,2]. This pristine geomorphological record can also be used to evidence topographic uplift related to the opening of an asthenospheric window below the continent [3]. Surrounding the Lago Buenos Aires (LBA), at ~46°30'S, there are well-preserved moraine deposits along with outwash and fluvial terraces that may be climatic counterparts. In the last decade, surface exposure dating of glacially-transported boulders of the youngest moraines has been carried out [2,3]. Nevertheless, constraints on the ages of the pre-LGM moraines are scarce, and surface exposure ages on terraces are completely missing. To chronologically constrain the oldest moraines (Deseado and Telken moraine systems) and the fluvial sequence, we applied ^{10}Be surface exposure dating on moraines preserved N and E of LBA, and on six outwash and fluvial terraces further east along the Rio Deseado.

Our results indicate: (1) Arid climate in Central Patagonia favoured the preservation of glacial and fluvial landforms allowing landscape and climate reconstructions back to 1 Ma. (2) Exposure ages from moraine boulders are scattered, but there is some consistency in the age ranges and the oldest ages obtained for the Deseado I and III moraines. They show an agreement between the Deseado and the Caracoles moraine system in the LBA and Lago Pueyrredon valleys, respectively. These data indicate a regionally significant glacial advance during MIS 16.

Fig. 1: *Photographs of Deseado moraine surfaces and moraine samples north of Lago Buenos Aires.*

(3) Exposure ages from the Telken moraine boulders are too young. They reflect boulder exhumation as a consequence of moraine degradation. However, dating of outwash cobbles on the associated Telken outwash yields much older exposure ages of ~1 Ma. (4) ^{10}Be exposure ages of 15 outwash and fluvial cobbles corresponding to the older terraces of the Rio Deseado yield stratigraphically consistent ages of 1.0-0.4 Ma. They mark a generalized uplift in Central Patagonia throughout the Quaternary.

[1]   A.S. Hein et al., Earth. Planet. Sci. Lett. 286 (2009) 184

[2]   M.R. Kaplan et al., Geol. Soc. Am. Bull. 116 (3/4) (2004) 308

[3]   B. Guillaume et al., Tectonics 28 (2009)

---

[1] *Geography, University of Bern*
[2] *IDEAN, University of Buenos Aires - CONICET*
[3] *Geography, Friedrich-Schiller-Universität Jena, Germany*

DECIPHERING LANDSCAPE EVOLUTION

Combined *in situ* ^{14}C-^{10}Be analysis in Earth surface sciences

K. Hippe

Reconstructing Quaternary landscape evolution today frequently builds upon cosmogenic-nuclide surface exposure dating. However, the study of complex surface exposure chronologies on timescales of 10^2-10^4 years remains challenging with the commonly used long-lived radionuclides (^{10}Be, ^{26}Al, ^{36}Cl). In glacial settings, key points are the inheritance of nuclides accumulated in a rock surface during a previous exposure episode and (partial) shielding of a rock surface after the main deglaciation event, e.g. during phases of glacier readvance. Combining the short-lived *in situ* cosmogenic ^{14}C isotope with ^{10}Be dating provides a valuable approach to resolve and quantify complex exposure histories and burial episodes within Lateglacial and Holocene timescales. The first studies applying the *in situ* ^{14}C-^{10}Be pair have demonstrated the great benefit from *in situ* ^{14}C analysis for unravelling complex glacier chronologies in various glacial environments worldwide (Fig. 1).

Although glacially modified landscapes have been a key target for *in situ* ^{14}C-^{10}Be dating, the short half-life of ^{14}C introduces various applications in sedimentary systems to quantify the processes in Earth surface development and landscape change. First studies have highlighted the capacity of combined *in situ* ^{14}C-^{10}Be analysis to quantify sediment transfer times in fluvial catchments, to constrain changes in surface erosion dynamics, or to date the formation of young sedimentary deposits (summarized in [1]). To exploit the full potential, however, future development needs to advance the current analytical techniques in order to increase the reliability of *in situ* ^{14}C analyses. The most relevant tasks are i) to improve the analytical reproducibility, ii) reduce the background level, and iii) to refine the *in situ* ^{14}C production rate calibration, notably for muonic production at depth. With these improvements,

the combination of *in situ* ^{14}C analysis with one or more long-lived nuclides could open up further novel opportunities for the use of cosmogenic nuclides to decipher the complex history of landscape development. Most importantly, simplified and more reliable analytical procedures should lead to a greater accessibility of researchers to *in situ* ^{14}C analyses, which is a prerequisite for an advanced use of *in situ* ^{14}C in the future.

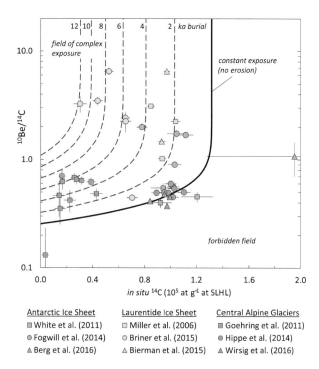

Antarctic Ice Sheet	Laurentide Ice Sheet	Central Alpine Glaciers
▫ White et al. (2011)	▫ Miller et al. (2006)	▪ Goehring et al. (2011)
⊙ Fogwill et al. (2014)	○ Briner et al. (2015)	● Hippe et al. (2014)
△ Berg et al. (2016)	△ Bierman et al. (2015)	▲ Wirsig et al. (2016)

Fig. 1: *Two-nuclide diagram of in situ ^{14}C vs. ^{10}Be/^{14}C summarizing published paired in situ ^{14}C-^{10}Be analyses from glacial settings. References are given in [1].*

[1]   K. Hippe, Quat. Sci. Rev. 173 (2017) 1

GROWTH OF RELIEF IN THE CENTRAL MENDERES MASSIF

Pattern of erosion revealed by cosmogenic ^{10}Be

C. Heineke[1], R. Hetzel[1], C. Akal[2], M. Christl

The central Menderes Massif in Western Turkey has undergone marked continental extension since the Miocene (e.g. [1]) and consists of two mountain ranges, which are bordered by the active Gediz and Büyük Menderes grabens (Fig. 1). To decipher how topographic relief evolves during continental extension, we have determined local erosion rates for ridge crests and spatially-averaged ^{10}Be erosion rates for catchments across the entire massif (Fig. 1, 2).

Fig. 2: *Exemplary catchment and ridge crest sampling site in the central Menderes Massif.*

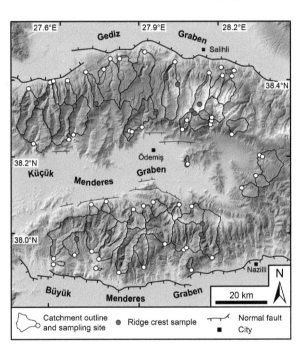

Fig. 1: *Map of the central Menderes Massif. Sampling sites are marked with red (ridge crests) and white dots (catchments).*

Samples collected from mountain ridges consisted of 1500-2500 bedrock clasts from regolith-mantled parts of the crests, to ensure that the local erosion rates are representative. The ridge crests erode at rates of 50-90 mm/kyr, which are 2-5 times lower than erosion rates for the adjacent catchments (i.e. 100-450 mm/kyr).

The relatively slower erosion of ridge crests as compared to the catchments indicates that local relief grows at rates of 50-200 and 150-350 m/Myr in the northern and southern mountain range, respectively. We interpret the increase in relief to be caused by ongoing normal faulting on the graben-bounding faults, which causes footwall uplift and concomitant river incision.

[1]   A. Wölfler et al., Tectonophysics 717 (2017) 585

[1] *Geology and Paleontology, University of Münster, Germany*
[2] *Geological Engineering, University of Izmir, Turkey*

QUANTIFYING CHEMICAL VS. MECHANICAL DENUDATION

Erosion rates and mechanisms in different lithologies, Crete, Greece

R. Ott[1], S. Gallen[1], S. Ivy-Ochs, M. Christl, C. Vockenhuber, N. Haghipour[1], S. Willett[1]

The Cretan landscape is dominated by high carbonate massifs defining the backbone of the island. The steep and high topography on the island is usually associated with carbonate units, whereas mountains in areas with clastic units are smaller and have less steep slopes (Fig. 1). On the eastern part of the island, sequences of large marine terraces cut into carbonates were shown to be of Pliocene age and have not been inundated during their Quaternary emergence, illustrating the resistance of the carbonate bedrock (Fig. 2). From these qualitative observations, it is assumed that under Mediterranean climate conditions the Mesozoic and Paleogene carbonates are more resistant to denudation than their partially metamorphosed clastic counterparts.

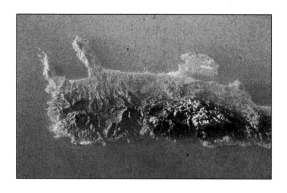

Fig. 1: *Carbonate massif (lower right corner) in western Crete surrounded by clastic units with lower topography.*

In order to quantify any differences in erosion rates, we collected sediments from the active streams of 15 catchments draining the phyllite-quartzite unit and four catchments draining the carbonate Plattenkalk and Tripalion units. Erosion rates were derived from [10]Be and [36]Cl measurements of alluvial sediments in clastic and carbonate catchments, respectively.

Apart from the difference in erodibility of different lithologies, we aim quantifying the proportion of chemical versus mechanical denudation at catchment scale. The ratio between both is crucial for a mass balance at orogenic scale, since the mass lost due to dissolution is permanent.

Fig. 2: *Pliocene marine terraces cut in carbonates showing little erosion.*

We collected seven samples of surface water, in conjunction with 65 water samples collected by Greek government agencies between the years 2000 and 2014, to quantify the dissolution rate of carbonates. This rate was then compared to a total weathering rate derived from [36]Cl measurements of alluvial carbonate sediments. First results show that denudation rates are around ~0.1 mm/a in the clastic units and ~0.13 mm/a in the carbonates. Calculated carbonate dissolution rates are ~0.6 mm/a. Therefore, only about half of the carbonate surface lowering can be attributed to dissolution. The higher denudation rates in the carbonates show that their topographic prominence is not solely a difference in erodibility but might be related to a tectonic and/or isostatic rebound origin.

---

[1] *Geology, ETHZ*

SPATIAL VARIABILITY EROSION RATES IN NW IRAN

^{10}Be concentrations in river sands from Ghezel-Ozan

A. Kaveh Firouz[1], J.P. Burg[1], N. Haghipour[1], S.K. Mandal[2], R. Elyaszadeh[3], M. Christl

Active convergence between the Arabian and Eurasian plates formed mountainous regions in NW-Iran. The role of active tectonics, climate and lithology in modulating this landscape remains little explored. We measured ^{10}Be concentrations in river sands of the Ghezel-Ozan catchment (Fig. 1) to determine catchment-averaged erosion rates and to test the hypothesis that active tectonics and exhumation of crystalline basement rocks led to transience in the landscape. The upper, middle and lower parts of the Ghezel-Ozan basin drain three tectono-stratigraphic zones with different lithology and relief. Moreover, historical and instrumental seismicity indicates fault activity in both the middle (central Iran) and lower (west Alborz zones) catchment (Fig. 2).

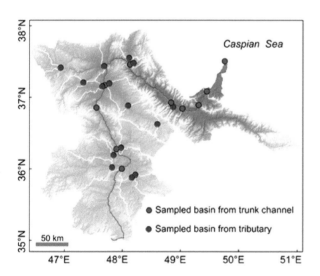

Fig. 1: *Sampled basins from trunk channel and tributaries of Ghezel-Ozan basin.*

We analysed channel metrics using digital elevation models and climatic parameters to explore their relationship with the millennial-scale erosion rates. The analyses along the catchment show no correlation between erosion rate and rainfall amount but correlation

a between channel steepness (ksn) and erosion rates in lower catchment.

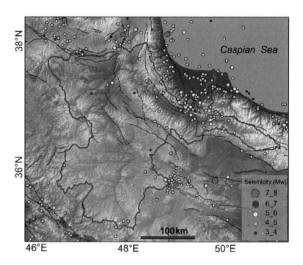

Fig. 2: *Topographic overview, major active faults (black lines), seismicity (colored circles), and tectonostratigraphic zones (colored area) along the studied Ghezel-Ozan basin.*

We found a two-fold difference in erosion rates between trunk channel and tributaries. This suggests a transient state and higher erosion rates in the lower catchment. The difference indicates that active tectonics in the lower catchment affect erosion rates under semi-arid climatic conditions.

---

[1] *Geology, ETHZ*
[2] *Geochemistry, Helmholtz Centre Potsdam, Germany*
[3] *Geological Survey of Tabriz, Iran*

SOIL FORMATION IN SUBTROPICAL TERRAIN

In-situ and meteoric ^{10}Be as tracers for soil development

V. Vanacker[1], J. Schoonejans[1], S. Opfergelt[1], B. Campforts[2], M. Christl

In super-humid mesothermic climates, subtropical soils develop under a lush vegetation cover (Fig. 1). These climatic conditions facilitate intense chemical weathering of the parent material and leaching of soluble minerals. Long-term soil development is a result of the interplay between soil formation and denudation, and their rates are not yet well constrained for subtropical regions.

Fig. 1: *Subtropical soils developed under semi-deciduous Araucaria forest (southern Brazil).*

We applied in-situ ^{10}Be and inventories of meteoric ^{10}Be to quantify long-term soil erosion and denudation in southern Brazil [1]. The acidic conditions of subtropical soils are less favorable for meteoric ^{10}Be retention, and we evaluated the potential loss of meteoric ^{10}Be in the soil system. The stable isotope ^{9}Be released upon weathering of parent material was used to track vertical redistribution of Be in the soil material.

Our data from three pits show incomplete retention of meteoric ^{10}Be in the soil system. The depth variation in the exchangeable, reactive and residual ^{9}Be fractions show that Be is mobilized within the soil profile by the amorphous oxy-hydroxide and crystalline oxide fractions. Based on the strong agreement in depth-variation of the reactive ^{9}Be and meteoric ^{10}Be concentrations (Fig. 2), we used the reactive ^{9}Be mass losses to correct for incomplete retention of meteoric ^{10}Be.

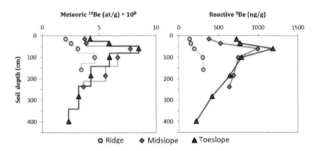

Fig. 2: *Depth variation of meteoric ^{10}Be and reactive ^{9}Be as measured in soil profiles along a toposequence.*

After correcting the inventories for incomplete retention of meteoric ^{10}Be, we derived surface erosion rates of ~5 mm/kyr for the slope convexities that agree well with in-situ ^{10}Be-derived denudation rates. Our results suggest that soil fluxes can be derived from meteoric ^{10}Be inventories in subtropical soils, when the meteoric ^{10}Be mobility is accounted for using e.g. differential mass balances.

[1] J. Schoonejans et al., Chem. Geol. 466 (2017) 380

[1] Earth and Life Institute, Université catholique de Louvain, Belgium
[2] Geo-Institute, Katholieke Universiteit Leuven, Belgium

SOIL EROSION IN A HUMMOCKY MORAINE LANDSCAPE

Detection of long-term rates using meteoric [10]Be

F. Calitri[12], M. Egli[1], M. Sommer[2], D. Brandová[1], M. Christl

Global change is expected to affect landscapes and mass fluxes from and into soils. Three sets of processes shape these changes: weathering, soil profile development and lateral redistribution of material. It is well known that these interact strongly. Periods with dominantly progressive soil forming processes (weathering) alternate with periods with dominantly regressive processes (erosion).

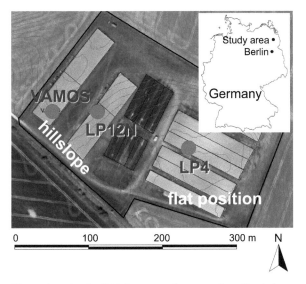

Fig. 1: *CarboZALF experiment in Dedelow, Uckermark, Germany.*

In the formerly glaciated areas of the lowlands of Uckermark, NE Germany, agriculture has strongly intensified since the early 1960s. The experimental site, CarboZALF (Fig. 1) is in a hummocky ground moraine landscape [1], characterised by kettle holes. Meteoric [10]Be has been measured along three different profiles in the CarboZALF site (Fig. 2) to evaluate long-term soil mass redistribution. This will be, in a next step, compared to short-term mass movements (caused by agricultural practice). In contrast to the hillslope sites 'LP12N' and 'Vamos', the site 'LP4' (Fig. 1), on a flat position, was hypothesised to exhibit no or negligible soil redistribution rates. Mass redistribution rates were calculated according to [2]. LP4 and VAMOS showed a slight accumulation of 0.37 and 0.08 t ha^{-1} y^{-1} respectively, and the LP12N site a strong erosion of 2.00 t ha^{-1} y^{-1}. The site VAMOS should, from a topographical point of view, be erosive. Erosion was, however, counterbalanced by the deposition of eroded (upslope) soil material about 2 ka BP (^{14}C-dated). This demonstrates that strong events of soil mass redistribution are not only a result of present-day activities (agriculture).

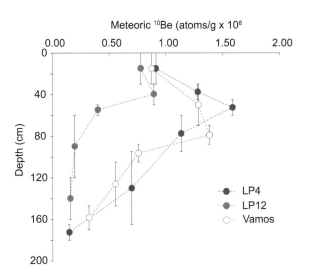

Fig. 2: *Depth profiles of meteoric [10]Be in soils.*

Further measurements of cosmogenic nuclides and fallout radionuclides ([239+240]Pu) will allow more detailed unravelling of the time-dependent processes of soil redistribution.

[1]   M. Sommer et al., Geoderma 145 (2008) 480
[2]   B. Zollinger et al., Earth Surf. Proc. Land 42 (2017) 814

[1] *Geography, University of Zurich*
[2] *ZALF, Müncheberg, Germany*

EARLY TO LATE PLEISTOCENE DEATH VALLEY FANS

Cosmogenic [10]Be and [26]Al in debris-flow deposits

M. Dühnforth[1], A. Densmore[2], S. Ivy-Ochs, P. Allen[3], P.W. Kubik

Debris-flow fans with depositional records over several 10^5 years may be useful archives for the understanding of fan construction by debris flows and post-depositional surface modification over long timescales. Reading these archives, however, requires that we establish the temporal and spatial pattern of debris-flow activity over time.

We used a combination of geomorphic mapping of fan surface characteristics, digital topographic analysis, and cosmogenic radionuclide dating using [10]Be and [26]Al to study the evolution of the Warm Springs fan on the west side of southern Death Valley, California [1]. The [10]Be concentrations yield dates that vary from (989±43) ka to (595±17) ka on the proximal fan and between (369±13) ka and (125±5) ka on distal fan surfaces. The interpretation of these results as true depositional ages though is complicated by high inheritance with a minimum of 65 ka measured at the catchment outlet and of at least 125 ka at the distal fan. Results from the [26]Al measurements suggest that most sample locations on the fan surfaces underwent simple exposure and were not affected by complex histories of burial and re-exposure. This implies that the Warm Springs fan is a relatively stable landform that underwent several 10^5 years of fan aggradation before fan head incision caused abandonment of the proximal and central fan surfaces and deposition continued on a younger unit at the distal fan. We show that the primary depositional debris-flow morphology is eliminated over a time scale of less than 10^5 years, which prevents the delineation of individual debris flows as well as the precise reconstruction of lateral shifts in deposition as we find it on younger debris-flow fans [2]. Secondary post-depositional processes control subsequent evolution of surface morphology

with the dissection of planar surfaces while smoothing of convex-up interfluves between incised channels continues through time.

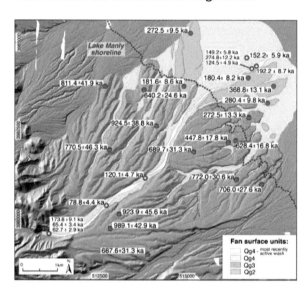

Fig. 1: *Relative and absolute chronology of fan units on the Warm Springs fan. Colored fan surfaces represent the relative chronology based on geomorphological criteria such as surface dissection and pavement development.*

[1] M. Dühnforth et al., Geomorph. 281 (2017) 53
[2] M. Dühnforth et al., JGR Earth Surface 112 (2007) F03S15

[1] *Earth and Environmental Science, LMU Munich, Germany*

[2] *Hazard and Risk Research and Department of Geography, Durham University, Durham, UK*

[3] *Earth Science and Engineering, Imperial College, London, UK*

SLIP-RATE OF THE OVACIK FAULT (TURKEY)

Surface exposure dating of boulders and depth profile dating with ^{36}Cl

C. Zabcı[1], T. Sançar[2], D. Tikhomirov[3], S. Ivy-Ochs, C. Vockenhuber, A.M. Friedrich[4], M. Yazıcı[1], N. Akçar[3]

The Anatolian *Scholle* is not only deformed along its main boundary structures, the North (NAF) and the East (EAF) Anatolian faults, but is also characterized by multiple internal NE-striking sinistral and NW-striking dextral faults. The Ovacık Fault is one of these sinistral structures, which is located at the easternmost part of Anatolia (Fig. 1).

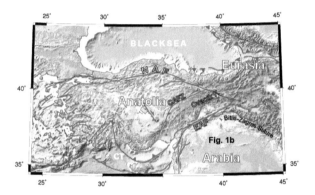

Fig. 1: *The major active tectonic elements of Turkey and the Ovacık Fault.*

We present the first slip rate estimates for the Ovacık Fault based on the morphochronology established by cosmogenic ^{36}Cl dating of offset fluvial deposits at the Köseler site (Fig. 2). We calculated different slip rate scenarios by using surface exposure or depth profile ages in assumption of three different cases, depending on whether the age of the upper tread, the lower tread, or all bounding surfaces were used in dating of terrace risers. The sharp displacement of the NF1/T2 riser displays a more reliable measurement than the broadly deflected NF1/T1 riser (Fig. 2).

The scatter of surface ages and variability of ^{36}Cl concentrations in depth profiles represent strong evidence for inheritance in alluvial fan and terrace deposits, whereas the width of the incised channel exceeds the magnitude of the sharp slip along the NF1/T2 riser.

Fig. 2: *The Köseler site. We measured offset at risers, separating the NF1 alluvial fan with the T1 (15-20 m) and T0 (27-37 m).*

Thus, we used the 15-20 m of horizontal slip of this offset feature and the modelled depth-profile age of the lower tread in order to calculate a slip rate of 2.5 (+0.7/-0.6) mm/a (2σ) for the Ovacık Fault. Our results together with other slip rate estimates along other active structures suggest plate boundary related intraplate deformation for Anatolia.

This study is under review in the journal of "Tectonics" and supported by TÜBİTAK grant no. 114Y227 and İTÜ-BAP MAB-2017-40586.

[1] Jeoloji Böl., İstanbul Teknik Üniversitesi, Turkey
[2] Jeoloji Böl., Munzur Üniversitesi, Turkey
[3] Geology, University of Bern
[4] Geology, LMU, Munich, Germany

PATTERNS OF LARGE LANDSLIDES IN THE ALPS

Comparing the timing of the Marocche di Dro events

S. Ivy-Ochs, S. Martin[1], P. Campedel[2], K. Hippe, C. Vockenhuber, M. Rigo[1], A. Viganò[2]

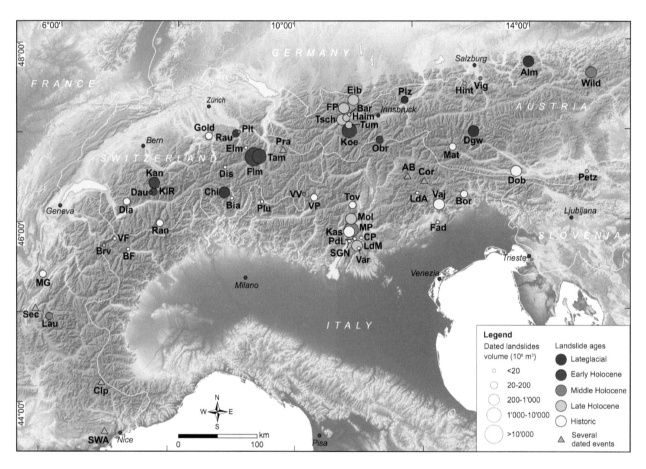

Based on geomorphology and ^{36}Cl exposure dating, the Marocche di Dro rock avalanches are divided into the Marocca Principale to the north and the Kas to the south. The former involved massive collapse of an estimated 1000×10^6 m^3 of limestone (5300±860) yr ago. During the Kas event 300×10^6 m^3 of rock detached (1080±160) yr ago.

The temporal and spatial distribution of large landslides in the Alps is shown in the figure (for details on site names and complete references see [1]). In the Alps, three periods of apparent enhanced rock slope failure have been recognized: 10-9 kyr, 5-3 kyr, and 2-1 kyr. No deposits of the first temporal cluster are found at Marocche di Dro. The age of Marocca Principale at (5300±860) yr broadly coincides with the recognized period of increased frequency of failure events at the transition from the middle to the late Holocene, (about 5-4 kyr) when a shift to a wetter, colder climate occurred. Nevertheless, a seismic trigger cannot be excluded.

[1] S. Ivy-Ochs et al., Quat. Sci. Rev. 169 (2017) 188

^{1} *Dipartimento di Geoscienze, Università di Padova, Italy*
^{2} *Servizio Geologico della Provincia Autonoma di Trento, Italy*

^{10}BE IN BLACK SEA SEDIMENTS

The Laschamp geomagnetic excursion as a global synchronization tool

M. Czymzik[1], N. Nowaczyk[2], H.W. Arz[1], R. Muscheler[3], M. Christl

Cosmogenic radionuclide production rate changes induced by varying geomagnetic field strength leave their signature in natural environmental archives. Detecting and aligning this signature provides the possibility for synchronizing records from different environmental archives and investigating the dynamics of climate variations in space and time, with minimized uncertainties in the relative timing [1].

Fig. 1: *Location of Black Sea sediment core 22-GC8 on the Archangelsky Ridge retrieved during 'R/V Meteor' cruise M72/5 in 2007.*

Variations of the cosmogenic radionuclide ^{10}Be concentration in Black Sea sediment core 22-GC8 (Fig. 1) provide a well-preserved record of the Laschamp geomagnetic excursion at 41000 a BP. Our time-series confirms the double-peak structure inferred from relative paleointensity changes reconstructed in sediments of the same archive, connected with an about 400-year full polarity reversal [2]. Synchronizing the Black Sea ^{10}Be time-series for the Laschamp event to a ^{10}Be record from central Greenland ice cores [3] using lag-correlation analysis reveals an offset between both records of 172 (+136/−123) years (max. r=0.68, p<0.01) (Fig. 2).

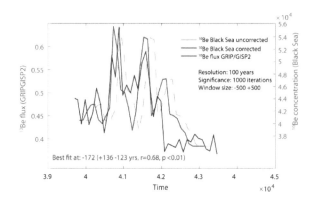

Fig. 2: *Preliminary synchronization of ^{10}Be from Black Sea sediments around the Laschamp geomagnetic excursion with ^{10}Be from central Greenland ice cores based on lag-correlation analyses. Significance of the correlations was calculated using 1000 iterations of a random phase test [4].*

This synchronization allows a direct comparison of proxy records reflecting Dansgaard-Oeschger events from the Black Sea and from Greenland and to investigate the evolution of these rapid climate changes in space and time.

[1] F. Adolphi & R. Muscheler, Clim. Past 12 (2016) 15
[2] N. Nowaczyk et al., EPSL 351-352 (2012) 54
[3] R. Muscheler et al., EPSL 219 (2004) 325
[4] W. Ebisuzaki, J. Clim. 10 (1997) 2147

[1] *Marine Geology, Leibniz Institute for Baltic Sea Research - IOW, Germany*
[2] *Climate Dynamics and Landscape Evolution, GFZ, Germany*
[3] *Quaternary Geology, Lund University, Sweden*

36Cl DATING IN THE NUBIAN SANDSTONE AQUIFER

Calibration with 81Kr ages and restoration of initial conditions

R. Purtschert[1], R. Ram[2], C. Vockenhuber

The Kurnub Group Nubian Sandstone Aquifer extends over large areas in the subsurface of the Sinai Peninsula (Egypt) and the Negev Desert (Israel). ^{81}Kr dating revealed groundwater ages in the range of 46-627 ka, far beyond former estimations gained by ^{14}C, with older groundwater ages in the southern Negev (300-630 ka) and younger ages in the north (50-330 ka). This is in contrast to the flow directions suggested by water levels in the aquifer (Fig. 1). The waters reveal also a distinct spatial (and temporal) pattern in the δ^{18}O-δ^2H composition, which may indicate recharge under different climatic conditions or from different recharge areas.

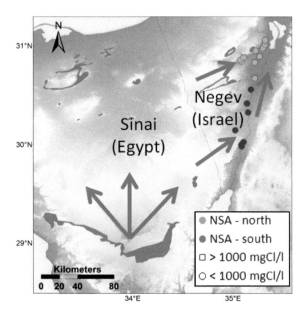

Fig. 1: *Recharge areas (in dark orange), main flow directions and sampled wells along the eastern margins of the NSA in the Negev and Arava Rift.*

The ^{81}Kr ages were used for estimating the initial Cl⁻ (Cl$_i$) concentrations, assuming an initial ^{36}Cl/Cl ratio of 90·10^{-15} and a subsurface secular equilibrium of 5·10^{-15} [1]. The calculated initial values correlate with the computed deuterium excess (d=8xδ^{18}O-δ^2H), with higher Cl$_i$ for lower

d-excess (Fig. 2). Hence, corrections were applied with differential Cl$_i$ values according to its computed d-excess values.

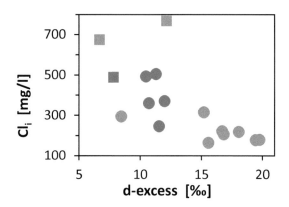

Fig. 2: *Calculated initial Cl⁻ vs. d-excess values.*

The corrected ^{36}Cl values and concluded water ages (Fig. 3) seem to be appropriate only for the northern samples.

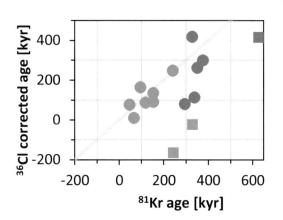

Fig. 3: *Corrected ^{36}Cl ages vs. ^{81}Kr ages.*

[1]   L.J. Patterson et al., Geochem. Geophys. Geosyst. 6 (2005)

[1] *CEP, Physics institute, University of Bern*
[2] *ZIWR Institute, Ben Gurion University, Israel*

THE GRIP ICE CORE ^{10}BE PROJECT

High-resolution ^{10}Be data to trace the solar 11-year cycle in the past

E. Anderberg[1], M. Christl, J. Beer[2], F. Adolphi[1,3], S. Bollhalder Lück, R. Muscheler[1]

The GRIP (Greenland Ice Core Project) ice core was drilled about 25 years ago and has since provided researchers with ample information about past climate changes and its forcings. ^{10}Be measurements were part of the GRIP project and a significant portion of the ice core was assigned to such measurements allowing for high-resolution investigations of past solar variability that influences the ^{10}Be production rate in the atmosphere.

First results showed the presence of the solar 11-year solar cycle, also known as the Schwabe cycle, during a short time interval (1780 – 2066 years B.C.) during the Holocene [1]. This cycle is known from the sunspot record (Fig. 1). However, the length of the Schwabe cycle is variable.

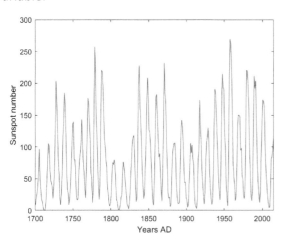

Fig. 1: *Recently revised international sunspot number from 1700 to now (source: WDC-SILSO, Royal Observatory of Belgium, Brussels, [2]).*

The aim of this project is to study the length of the Schwabe cycle using cosmogenic radionuclides beyond the reach of sunspot observations. Consequently, sampling of the GRIP ice core was repeated with a higher depth resolution compared to earlier measurements (27.5 cm for the period between 4100 and 7800

years BP instead of the normal 55 cm bag resolution). In collaboration with the French research group led by G. Raisbeck we have now completed the remaining measurements. Approximately half of the samples were prepared in Lund, Sweden and measured at ETH while the other half was prepared in Orsay and measured at the Tandetron in Gif sur Yvette or by the AMS group in Aix en Provence. This completed high-resolution ^{10}Be record will enable us to study the Schwabe cycle and its length much further back in time. Figure 2 shows the sampling resolution of the ^{10}Be data for the Holocene part of the GRIP project. The unprecedented resolution will allow gaining insights into solar variability in the past and its stability.

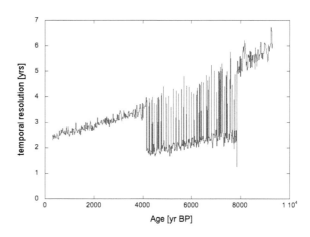

Fig. 2: *Sampling resolution of the GRIP ^{10}Be data. The period of higher sampling resolution is clearly visible.*

[1] F. Yiou et al. JGR 102 (1997) 783
[2] F. Clette et al. Space Sci. Rev. 286 (2014) 35

[1] *Dept. of Geology, Lund University, Sweden*
[2] *EAWAG, Dübendorf*
[3] *Climate and Environmental Physics, University of Bern*

ANTHROPOGENIC RADIONUCLIDES

Artificial ^{236}U and ^{129}I in the North Atlantic

Distribution of ^{129}I and ^{236}U in the Fram Strait

Presence of ^{236}U at the Canada Basin, Arctic Ocean

Is there a source of ^{236}U in the Baltic Sea?

^{236}U at the mouth of the Columbia River

Results of ^{129}I from the GO-SHIP P16N (2015) line

Simulating ^{236}U with a global ocean model

Measurement of ^{236}U in Celtic Sea sediment cores

Plutonium in drinking water reservoirs

Radionuclides in drinking water reservoirs

Analyses of proton-irradiated tantalum targets

Dating nuclear fuel with curium isotopes

ARTIFICIAL ^{236}U AND ^{129}I IN THE NORTH ATLANTIC

Tracing water masses along the GEOVIDE transect in spring 2014

M. Castrillejo[1], P. Masqué[1], J. Garcia-Orellana[1], N. Casacuberta, C. Vockenhuber, M. Christl

Nuclear reprocessing plants and atmospheric weapon tests have introduced artificial radionuclides into the oceans since the 1950s. These radionuclides can be used as tracers of water masses involved in the Atlantic Meridional Overturning Circulation (AMOC), which plays a major role in the forcing of Earth's climate. In this study we investigate the water mass transport pathways and mixing processes composing the AMOC. Long-lived ^{129}I ($T_{1/2}$ = 15.7 Ma) and ^{236}U ($T_{1/2}$ = 23.5 Ma) were measured in over 300 seawater samples collected along the GEOVIDE transect in spring 2014 (inset Fig. 1).

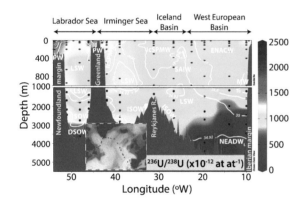

Fig. 1: *Distribution of ^{236}U/^{238}U atomic ratio along the GEOVIDE transect.*

The ^{236}U/^{238}U atomic ratios range from (40±20) x 10^{-12} to (2350±370) x 10^{-12} and from (0.20±0.20) x 10^7 to (256±4) x 10^7 at·kg^{-1} for ^{129}I concentrations. The ^{236}U/^{238}U atomic ratios (Fig. 1) and ^{129}I concentrations (not shown) increase from east (Portugal) to west (Canada). They are lowest near the seafloor in the West European Basin indicating mixing between deep Atlantic Water (AW) and older waters of Antarctic origin carrying almost no ^{236}U and ^{129}I (Fig. 1). A large presence of tracers is observed in Polar Waters (PWs) flowing over the Greenland Shelf and Newfoundland Shelf. The ^{129}I and ^{236}U will be used to study mixing

processes between AWs and PWs entering through the Canadian Archipelago and the Fram Strait from the Arctic Ocean.

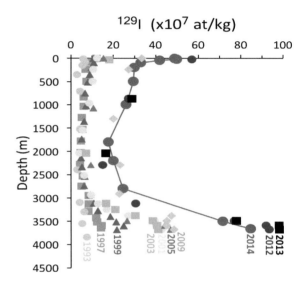

Fig. 2: *Concentrations of ^{129}I in the Labrador Sea for 1993-2013 [1-4] and 2014 (GEOVIDE).*

Tracer levels are also high in overflow waters (e.g. DSOW) ventilating the deep Labrador and Irminger Basins (Fig. 1). The ^{129}I concentrations in near-bottom flowing DSOW have increased over time due to the increased discharge rate of ^{129}I effluents from European reprocessing plants (Fig. 2). In a next step, the time series will be discussed in relation to recent changes in ocean circulation of the subpolar North Atlantic.

[1]   H. Edmonds et al., JGR Oc., 106 (2001)
[2]   J. Smith et al., JGR Oc. 110 (2005)
[3]   S. Orre et al., Env. Fluid. Mech. 10 (2010) 213
[4]   J. Smith et al., JGR Oc. 121 (2016)

---

[1] *Institut de Ciència I Tecnologia mbientals & Dept. de Física, Universitat Autònoma de Barcelona, Spain*

DISTRIBUTION OF ^{129}I AND ^{236}U IN THE FRAM STRAIT

Tracing water masses by ^{129}I and ^{236}U released from reprocessing plants

A.-M. Wefing, N. Casacuberta, M. Christl, C. Vockenhuber, M. R. van der Loeff[1]

The Fram Strait is a region of particular importance regarding water mass circulation, as it is the only deep gateway allowing for intermediate and deep water exchange between the North Atlantic and the Arctic Ocean. The long-lived artificial radionuclides ^{129}I and ^{236}U are known to be suitable tracers to study circulation patterns in the Nordic Seas and the Arctic Ocean due to their locally and timely constrained release by the two European nuclear reprocessing plants in Sellafield (UK) and La Hague (France).

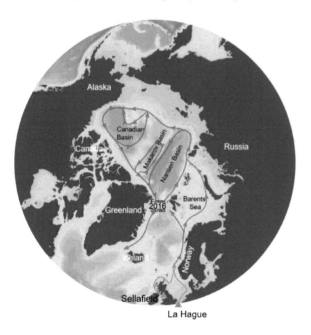

Fig. 1: *Circulation of Atlantic waters in the Arctic Ocean and pathways of ^{129}I and ^{236}U from Sellafield and La Hague. Yellow arrow shows the sampled transect in 2016.*

For the compilation of a first comprehensive dataset of ^{129}I and ^{236}U from the Fram Strait, about 300 water samples (19 deep profiles) were taken during the R/V Polarstern cruise PS100 "GRIFF" in 2016. Samples were pre-processed onboard and radionuclide concentrations were measured with the compact AMS system Tandy.

Lowest concentrations of radionuclides are found in deep Atlantic waters flowing into the Arctic Ocean (Fig. 2). Highest concentrations are observed in the surface waters of the East Greenland Current (EGC; outflow from the Arctic Ocean), which are nearly twice as high as concentrations measured in the West Spitsbergen Current (WSC; inflow to the Arctic Ocean). High concentrations are represent-tative of Atlantic waters carrying the signal of the reprocessing plants, presumably indicating the return flow of Atlantic waters that have circulated through the Arctic Ocean. Additional inputs could be of riverine origin, being this a subject for further investigations.

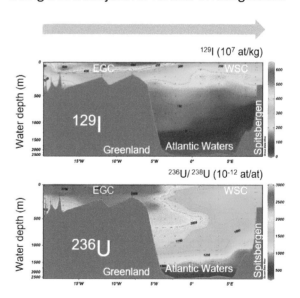

Fig. 2: *^{129}I concentration and ^{236}U/^{238}U ratio in a west to east section through the Fram Strait (yellow arrow in Fig. 1).*

[1] *Alfred Wegener Institute, Germany*

PRESENCE OF ^{236}U IN THE CANADA BASIN, ARCTIC OCEAN

First results from the GEOTRACES US Arctic expedition GN01 (2015)

E. Chamizo[1], N. Casacuberta, M. Christl, T. Kenna[2], M. López-Lora[1,3], P. Masqué[4,5], M. Villa[6]

During the US Arctic expedition in 2015, about 170 samples (4 deep and 20 shallow profiles) were collected for ^{236}U analysis in the Canada Basin (Fig. 1). The aim was to assess the presence of anthropogenic ^{236}U in an area of the Arctic Ocean not explored before, completing a transect across the Eurasian and the Amerasian Basins [1]. Uranium was pre-concentrated at Lamont-Doherty Earth Observatory (LDEO) using 4-10 L samples. No ^{233}U spike was added to prevent ^{236}U contamination. Fe(OH)$_3$ precipitates were then chemically processed at the Centro Nacional de Aceleradores (CNA) for purification of U isotopes [2]. Most surface and shallow samples were kept at the CNA for the analysis on a 1 MV AMS system. Samples below 400 m depth were sent to the ETH-Zurich to be analysed on the 600 kV Tandy AMS facility.

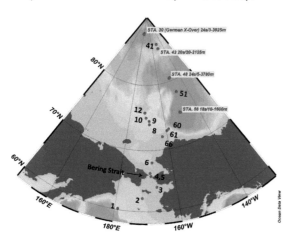

Fig. 1: *Location of the stations sampled during the US expedition.*

The results obtained so far show that average ^{236}U/^{238}U atom ratios (AR) in surface waters at the Bering Strait are 860 × 10^{-12} (St. 3, 5), in agreement with reported data from the North Pacific Ocean, mostly influenced by global fallout [3]. Entering the Canada Basin, ^{236}U/^{238}U ratios in surface samples reach values above 1000 × 10^{-12} (St. 8). The lowest ^{236}U/^{238}U ratios (~10^{-12}) were measured in the abyssal samples from St. 48, probably dominated by natural ^{236}U (Fig. 2). The remaining dataset will be analyzed during 2018. Additionally, ^{237}Np analysis will be performed in two deep profiles at the LDEO by ICPMS and at CNA. The joint study including both ^{237}Np and ^{236}U and the dataset of the US expedition together with the German cruise (PS94, 2015) shall be instrumental in understanding the circulation patterns in the Arctic Ocean and beyond.

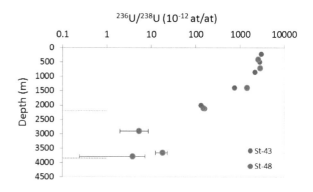

Fig. 2: *Preliminary ^{236}U/^{238}U results for the two depth profiles in stations 43 and 48. Bottom depths marked by dashed lines.*

[1] N. Casacuberta et al., EPSL, 440 (2016) 127
[2] M. Villa et al., Chem. Geo. (in preparation)
[3] R. Eigl et al., JER, 169-170 (2017) 70

[1]*Centro Nacional de Aceleradores, Sevilla, Spain*
[2]*Lamont-Doherty Earth Observatory. Columbia University. USA*
[3]*Applied Physics, Universidad de Sevilla, Spain*
[4]*Centre for Marine Ecosystems Research, Edith Cowan University, Australia*
[5]*Física & Ciència i Tecnologia Ambientals, Universitat Autònoma de Barcelona, Spain*
[6]*Applied Physics, Universidad de Sevilla, Spain*

IS THERE A SOURCE OF ^{236}U IN THE BALTIC SEA?

New U data from the North Sea and comparison with literature say yes

M. Christl, N. Casacuberta, J. Lachner, J. Herrmann[1], H.-A. Synal

In this study we produced new ^{236}U and ^{238}U data from seawater sampled in the North Sea in 2010 [1]. The North Sea receives considerable input of anthropogenic radionuclides from nuclear reprocessing facilities located in La Hague (France) and Sellafield (Great Britain). It therefore represents an important source region for oceanographic tracer studies using the transient signal of, for example, anthropogenic ^{236}U. For this purpose, a proper knowledge of the sources of ^{236}U is an essential prerequisite.

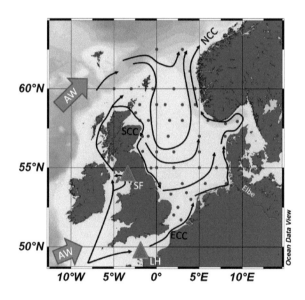

Fig. 1: *Map of the sampling region and location of the water samples (blue dots) together with the main water currents (arrows). The red triangles indicate the location of Sellafield (SF) and La Hague (LH), respectively.*

The new ^{236}U data set covers the transition regions of the North Sea to the Atlantic Ocean, to the Baltic Sea, and upstream the Elbe River and therefore allows to check for additional sources of ^{236}U entering the North Sea via Baltic Sea or the Elbe River (Fig. 1).

Our results show that both ^{236}U concentrations and ^{236}U/^{238}U ratios in surface waters of the North Sea can be explained by simple binary mixing models (Fig. 2) implying that ^{236}U behaves as conservatively as ^{238}U in seawater. Our results further show that the input of ^{236}U by the Elbe River is negligible, while there has been (or still is) a significant input of ^{236}U via the Baltic Sea (data from ref. [2]).

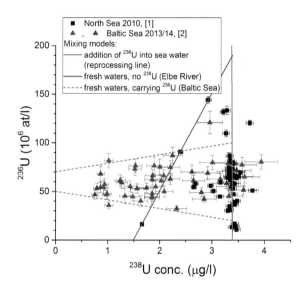

Fig. 2: *Binary mixing models indicating the addition of ^{236}U to seawater (red) and the possible input of ^{236}U with fresh water (blue: no additional ^{236}U, purple: with ^{236}U).*

The results of the mixing models suggest that this yet unidentified source of ^{236}U is most probably supplied by fresh water input via the Baltic Sea. More data from that region is needed to further constrain the additional ^{236}U input.

[1]   M. Christl et al., ES&T 12153 (2017) 51
[2]   J. Quiao et al., ES&T 6876 (2017) 51

[1] *BSH Hamburg, Germany*

^{236}U AT THE MOUTH OF THE COLUMBIA RIVER

Columbia River as a source of ^{236}U to the Pacific Ocean

M. Christl, N. Casacuberta, K. Buesseler[1], H.-A. Synal

The Hanford site was set up in 1943 as part of the Manhattan Project. It is located at the riverbank of the Columbia River (Washington) and consisted of several nuclear reactors and plutonium processing units essentially used for US nuclear weapons production. After the end of the cold war weapons production reactors were decommissioned leaving behind 200000 m^3 of liquid high-level radioactive waste in several storage tanks, 710000 m^3 of solid radioactive waste and 520 km^2 of contaminated groundwater beneath the site [1]. Hanford is currently the most contaminated nuclear site in the US and is the focus of the nation's largest environmental cleanup [1].

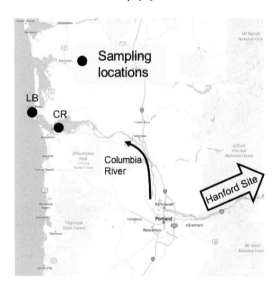

Fig. 1: Map of the sampling locations at the mouth of Columbia River (CR) and at the coastal Pacific Ocean (LB). The thin arrow indicates Columbia River. Hanford site is outside the map, upstream the river (large arrow).

Two water samples were collected in 2017 to check whether the Columbia River still represents a source of ^{236}U (and ^{129}I) for the Northeastern Pacific Ocean. The samples were taken at the coastal Pacific Ocean (Long Beach,

LB) and a few kilometers upstream (CR). At these locations, the water samples represent a mixing between Pacific Ocean water and Columbia River water. To estimate the ^{236}U concentration in pure river water the ^{238}U concentration was used as a proxy for salinity and simple binary mixing was assumed between seawater and the Columbia River source (Fig. 2).

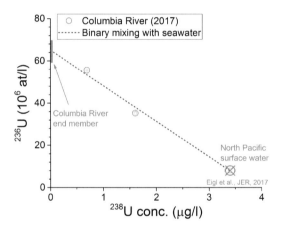

Fig. 2: Binary mixing model (blue line) indicating the addition of ^{236}U with fresh water to Pacific Ocean surface water.

Our results show that ^{236}U and ^{238}U concentrations follow a mixing line that represents simple binary mixing of Pacific Ocean surface water with a fresh water source (no or very low ^{238}U) carrying a ^{236}U concentration of 6-7 x 10^7 at/l. Assuming an annual water discharge of 2 x 10^{11} m^3/yr with a constant ^{236}U concentration results in an annual input of 6 g ^{236}U into the Pacific Ocean which is small compared to the global fallout inventory of about 1000 kg.

[1] https://en.wikipedia.org/wiki/Hanford_Site

[1] *WHOI, Woods Hole, USA*

RESULTS OF ^{129}I FROM THE GO-SHIP P16N (2015) LINE

Is Hanford Site a major source of ^{129}I to the Pacific Ocean?

N. Casacuberta, C. Vockenhuber, M. Christl, A.-M. Wefing, N. Chu[1], A.M. MacDonald[1], K.O. Buesseler[1]

During the oceanographic expedition of GO-SHIP P16N (2015, including a transect through the Alaska gyre) more than 400 seawater samples were collected for the analysis of artificial radionuclides (Fig. 1). The primary goal was to study the influence and dispersion of the Fukushima-derived releases that occurred during the nuclear accident in 2011.

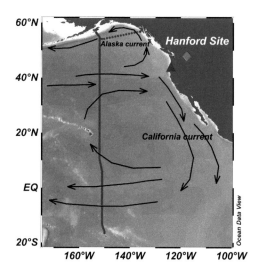

Fig. 1: *Map showing the location of the sampled stations in the open ocean (blue dots) and at the mouth of the Columbia River (blue triangle). Black arrows represent ocean currents.*

^{129}I from these transects ranged from (0.3 ± 0.2) x 10^7 at·kg^{-1} to (120 ± 2) x 10^7 at·kg^{-1}. Values in the lower range (up to $2 \cdot 10^7$ at·kg^{-1}) were observed in equatorial latitudes and correspond to the weapon test signal introduced in the 1950s-1960s from atmospheric deposition. Higher values ($>25 \cdot 10^7$ at·kg^{-1}) were observed in the northern part of the P16N section, particularly in the Alaska current (Fig. 2). However, the distribution of ^{129}I concentrations had no correlation to the unequivocal signal of Fukushima-derived ^{134}Cs measured in these samples clearly pointing to an alternative source of ^{129}I for the northeast Pacific Ocean. Could the

Hanford Site be a major source of ^{129}I? Two samples taken in 2017 at the mouth of the Columbia River (Astoria) and Long Beach (Fig. 1, blue triangle) showed ^{129}I concentrations of (8.3 ± 0.2) x 10^7 at·kg^{-1} and (6.7 ± 0.2) x 10^7 at·kg^{-1}, respectively. These values are consistent with those reported in 1996 [1] but cannot explain the high concentrations observed in the Alaska current in 2015, supporting the idea of having a potential source of nuclear waste upstream Columbia River. Several leakages from the storage tanks at the Hanford site have been reported in recent years, but our ^{129}I results suggest that this point source has not been constant or decreasing with time.

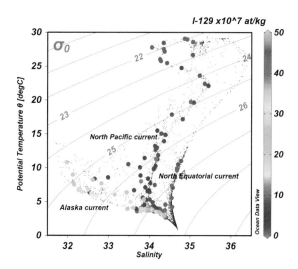

Fig. 2: *T-S plot with superimposed ^{129}I concentrations (in coloured dots).*

Further studies will be performed in this area to better constrain the sources of ^{129}I to the northeast Pacific Ocean.

[1] J.E. Moran et al., Water Resour. Res. 38 (2002) 24

---

[1] *WHOI, Woods Hole, USA*

SIMULATING ^{236}U WITH A GLOBAL OCEAN MODEL

First steps of implementing ^{236}U in the Community Earth System Model

N. Casacuberta, M. Christl, A.-M. Wefing, S. Yang[1], N. Gruber[1]

Several experimental studies have already proven the potential of using ^{236}U as a new oceanographic tracer [1]. This is especially relevant in the North Atlantic and Arctic Ocean, where the combination of ^{236}U with other artificial radionuclides from the two European Reprocessing Plants (RP) can help to constrain sources of water masses and ultimately be used to calculate tracer ages in the North Atlantic and Arctic Ocean. Yet, to fully exploit the potential of this tracer we need to better constrain the sources of ^{236}U to the marine environment. To achieve this, we made use of the ocean global model: Community Earth System Model (CESM).

First, we forced the global ocean model to the output of an atmospheric model (ECHAM) that simulated the atmospheric ^{236}U input from nuclear weapons testing (Fig. 1). Second, we introduced the two point-like sources of Sellafield and La Hague, simulating the reprocessing plant signal of ^{236}U and ^{129}I. The ^{236}U input function from Sellafield and La Hague was taken from Christl et al. [2].

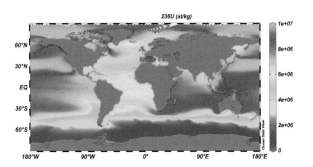

Fig. 1: *Simulated ^{236}U surface concentrations after forcing the ocean global model CESM to the output of the atmospheric model ECHAM.*

The main finding of this exercise was that the global ocean model CESM could reproduce the dispersion of weapon test ^{236}U in the oceans, confirming the estimated amount of approximately 1000 kg of ^{236}U. However, the global ocean model could not reproduce the distribution of ^{236}U and ^{129}I from the two European reprocessing plants beyond the North Sea domain due to the lack of spatial resolution (Fig. 2). Consequently, the model could not be applied in the North Atlantic and Arctic Ocean domain, where most of the ^{236}U and ^{129}I from reprocessing plants is being transported through more advective structures, such as the Norwegian Coastal Current and the Arctic Boundary Current.

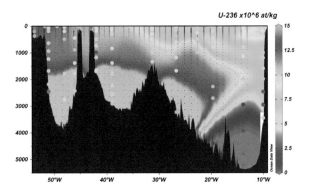

Fig. 2: *GEOVIDE section showing the observed ^{236}U concentrations superimposed on the model output. Inset map shows the section location.*

Next steps will involve the use of a regional ocean model (NAOSIM) run by the Alfred Wegener Institute. New and existing data will be combined with simulations in a 3-dimensional ice-model (NAOSIM) to understand the fate of the source branches of Atlantic Waters in the Arctic Ocean.

[1] N. Casacuberta et al. EPSL 440 (2016) 127

[2] M. Christl et al. JGR 120 (2015) 7282

[1] *Environmental Physics, ETHZ*

MEASUREMENT OF ^{236}U IN CELTIC SEA SEDIMENT CORES

Influence of Sellafield NFRP and comparison with other radionuclides

M. Villa-Alfageme[1], N.Casacuberta, E. Chamizo[2], S. Hurtado-Bermúdez[1] M. Christl

Two seafloor sediment cores from the Celtic Sea, South of Sellafield Nuclear Fuel Reprocessing Plant (NFRP), operating since the 60's, were processed and analyzed for ^{236}U at ETH-LIP. Artificial ^{137}Cs and ^{241}Am as well as natural ^{210}Pb, ^{40}K, ^{234}Th - to be used as possible chronological markers - were measured at University of Sevilla by gamma-spectrometry. Site A is sandy mud and Site I is muddy sand (~350 m depth) (Fig. 1, green stars). ^{236}U results are compared with the few measurements published up to date in cores of the N. Atlantic Ocean (~4500 m depth) [1] and Japan Sea (~2700 m depth) [2].

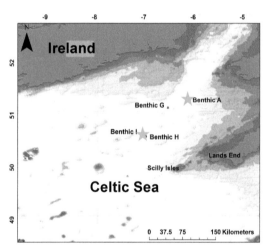

Fig. 1: *Map of the Celtic Sea and location of the cores A and I.*

^{236}U concentrations are lower in the core A (sandy) compared to core I (muddy). ^{236}U is detected deeper than expected (down to 14 cm). Since sedimentation rate at the Celtic Sea is approximately 0.08 cm yr^{-1}, this points out to a strong turbation.

Inventories, ranged from 1.32 x 10^{12} at/m^2 to 1.97 x 10^{12} at/m^2 in A and I; one order of magnitude higher than the ones from PAP site and Japan Sea, exclusively affected by fallout, pointing out a possible influence of the discharges from Sellafield Reprocessing Plant. However, direct comparison of the data is not recommended due to the incomplete collection of sedimented ^{236}U in A and I cores and contrasting depths and sediment composition.

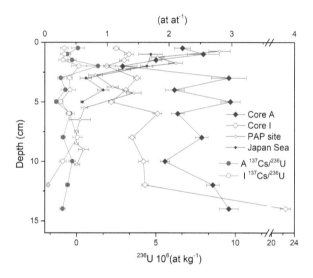

Fig. 2: ^{236}U *and* ^{137}Cs/^{236}U *ratios in sediments cores A and I, PAP [1] site and Japan Sea [2].*

The relatively constant ^{137}Cs/^{236}U ratios measured in most of the depth are in agreement with the very similar ^{137}Cs and ^{236}U input functions from Sellafield. In addition ^{137}Cs/^{236}U ratios are in agreement in both A and I cores and inventory ^{137}Cs/^{236}U ratios ranged from (0.14±0.05 to 0.20±0.07) for A and I, respectively.

[1]   M. Villa-Alfageme et al., JER (2017) In press
[2]   A. Sakaguchi et al., EPSL 165 (2012) 333

[1] *University of Sevilla, Dpto. Física Aplicada II, Spain*
[2] *University of Sevilla-CSIC-JA. CNA, Spain*

PLUTONIUM IN DRINKING WATER RESERVOIRES

Migration of Plutonium into North German groundwater

S. Bister[1], S. Pottgießer[1], B. Riebe[1] , C. Walther[1], M. Christl, H.-A. Synal

Water is an essential resource for life and is exposed to different environmental pollutants originating from human activities. Radionuclides entering aquatic ecosystems, regardless of whether man-made or naturally occurring, are all considered as contaminants. Within the scope of a joint project ('TransAqua') funded by the German Federal Ministry of Education and Research (BMBF / 02 NUK 030D), open questions from a radioecology point of view were addressed.

As a part of this project, samples of two drinking water abstraction areas ('Fuhrberger Feld' aquifer, Harz mountain reservoirs) in northern Germany were analyzed. Different environmental media were investigated, including stagnant and streaming water bodies, ground waters from different depths, as well as topsoils and forest humus layers. Plutonium isotopes were investigated as an example of man-made radionuclides in addition to other radionuclides like ^{3}H, ^{14}C, ^{90}Sr, ^{137}Cs, and ^{129}I. As we expected concentrations to be very low, samples of high volumes were required, which in turn were accompanied by complex sample preparation [1].

Fig. 1: *Sampling of surface-near groundwater at 'Fuhrberger Feld'.*

Plutonium could be detected in four samples, all of which had been collected from forest humus layers.

Fig. 2: *Sampling of the forest humus layer at 'Fuhrberger Feld'.*

These results confirm that plutonium is generally poorly soluble and has a low tendency to migrate. Massic activities for ^{239}Pu + ^{240}Pu found in the humus were in the range of (0.14±0.01 Bq/kg) to (1.26±0.03 Bq/kg).

Tab. 1: *Results of the most relevant samples.*

Sample	$^{240}Pu/^{239}Pu$
T3-forest _humus	0.19 ± 0.007
T3-forest _humus II	0.18 ± 0.007
SW1-forest _humus	0.18 ± 0.016
SW1-forest _humus II	0.19 ± 0.007

The isotope ratios provide information about the origin of the contamination. Global fallout is indicated by ^{240}Pu/^{239}Pu ratios of 0.18±0.01 [2]. Results reveal that the contamination of our samples is mainly caused by global fallout.

[1] S. Schneider, et al., Appl. Geochem. 85 (2017) 194

[2] J. Kelley, et al., Sci. Total. Environ. 237/238 (1999) 483

[1] *Radioecology and Radiation Protection, University of Hannover, Germany*

RADIONUCLIDES IN DRINKING WATER RESERVOIRES

Sensitivity of reservoirs to input of man-made radionuclides

B. Riebe [1], S. Bister[1], C. Walther[1], C. Vockenhuber, H-A. Synal

Water is exposed to different environmental pollutants originating from human activities, including radionuclides. In a project aiming to assess the sensitivity of an unconfined aquifer in Northern Germany (Fuhrberger Feld), which is used as a drinking water reservoir, we analysed water and soil samples with regard to ^{3}H, ^{14}C, ^{90}Sr, ^{137}Cs, Pu isotopes, and ^{129}I.

Ground water samples were drawn from multilevel-wells at different depths (3 to 17 m) aligned in the direction of ground water flow (Fig. 1). Additionally, samples from adjacent rivers and ponds were collected, as well as samples from forest and farmland soils. For analysis of ^{129}I the compact AMS system TANDY is used, ^{127}I concentrations are determined by inductively coupled plasma mass spectrometry.

Fig. 1: *Drawing ground water samples from a multi-level well in the 'Fuhrberger Feld' .*

Despite the fact that most of the ^{129}I input from the atmosphere is retained by top soils and humus layers ('litter'), respectively, results of the analyses revealed that ^{129}I activity concentrations in ground water samples from Fuhrberger Feld (FF) were about one order of magnitude higher than those from ground water samples from other aquifers from Lower Saxony (LS) (Fig 2.). For FF activity concentrations of ^{129}I

varied between 6.2×10^{-8} and 5.2×10^{-7} Bq kg^{-1}, and therefore lie within the range of values determined for surface water samples from Lower Saxony. In comparison, the ^{129}I activity concentrations of ground water samples from confined aquifers from the same region were in a range of 8.5×10^{-9} to 7.4×10^{-8} Bq kg^{-1}.

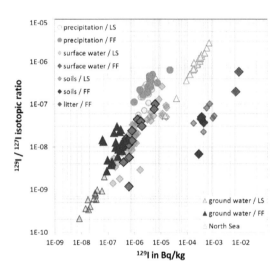

Fig. 2: : *^{129}I/^{127}I isotopic ratios versus ^{129}I for water and soil samples from Fuhrberger Feld (FF) in comparison to other environmental compartments in Lower Saxony (LS), and North Sea water from [1].*

The same is true for the ^{129}I/^{127}I ratio. Values of ground water samples from FF vary between 2.7×10^{-9} and 3.0×10^{-8}, which is one order of magnitude higher than the ^{129}I/^{127}I ratio of LS ground water samples. This indicates that ^{129}I from the atmosphere has already reached the Fuhrberger Feld aquifer.

[1]   R. Michel et al., Sci. Tot. Environ. 419 (2012) 151

[1] Institute for Radioecology and Radiation Protection, Leibniz University Hannover, Germany

ANALYSES OF PROTON-IRRADIATED TANTALUM TARGETS

Cross section measurements of ^{36}Cl

Z. Talip[1], C. Vockenhuber, J.–C. David[2], D. Schumann[1]

Reliable cross section data of the radionuclides produced in target materials used in spallation neutron sources (SNF) and accelerator driven systems (ADS) are essential for the safety assessment of these facilities. ^{36}Cl ($T_{1/2}$: 301000 yr) is of special interest for the decommissioning and especially for assessing accident scenarios due to its long half-life, volatility and high mobility in the geosphere.

Separation of Cl from proton irradiated Ta targets was given in the previous report [1]. In this study, the production cross sections of the ^{36}Cl were determined, using the LIP TANDEM AMS facility.

Uncertainties due to the AMS measurements were between 2-17%. Variable blank results showed that cross contamination during sample preparation should be taken into consideration. Therefore, blank corrections were performed for all the samples. Only the data, which have values significantly higher than the blanks were used for the cross section calculations.

Cumulative cross sections of ^{36}Cl were determined with the following equation (^{36}Cl/^{35}Cl ratios were converted to ^{36}Cl/Cl ratios by applying the conversion factor 0.7577);

$$\sigma = \frac{R_s\, N}{N_t\, \phi\, t_i}$$

where R_s is the measured isotopic ratio in the sample, N is number of carrier atoms, t_i is the irradiation time, N_t is the number of target atoms (which is much smaller than N), φ is the proton flux density.

In addition, experimental cross section results were compared with INCL4++-ABLA07 code. Figure 1 shows the comparison of the experimental and theoretical cross section results. There is a good agreement for the high energy proton-irradiated samples (> 750 MeV), while experimental cross section results are higher for the low energy proton-irradiated samples compared with the theoretical predictions [2]. Previously, similar results were observed for ^{36}Cl in proton-irradiated Bi and Pb targets [3].

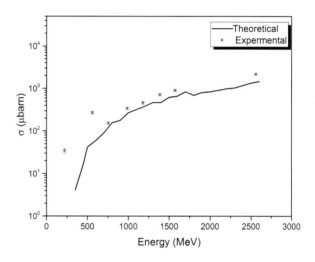

Fig. 1: *Experimental and theoretical cross section results for ^{36}Cl in proton-irradiated Ta.*

Further AMS measurements will be performed to determine the production cross sections of the ^{129}I and ^{36}Cl in a proton-irradiated W target, which will be used as a target material for the European Spallation Source.

[1]   Z. Talip et al., LIP annual report (2016) 92

[2]   Z. Talip et al., Anal. Chem. 89 24 (2017) 13541

[3]   D. Schumann et al., Nucl. Data Sheets 119 (2014) 288

[1] *PSI, Villigen*

[2] *Irfu, CEA, Universite Paris-Saclay, France*

DATING NUCLEAR FUEL WITH CURIUM ISOTOPES

Novel chronometry model to date nuclear fuel

N. Guérin[1], Z. Kazi[1], S. Burrell[1], A. Gagné[1], M. Totland[1], M. Christl

Illicit trafficking of nuclear material is a serious threat in the hands of criminal and terrorist organizations. It could be used to prepare radiological dispersive devices (RDD). When an illicit nuclear material is found, nuclear forensics analyses are conducted on the radioactive sample to determine its chemical/radiological composition, physical structure, and age since production or use. The information obtained is used to try to determine where and when the illicit material was produced, purified, enriched, irradiated, taken from regulatory control, if a crime has been committed and possibly provide evidence to bring charges. Current radiological chronometry techniques provide a date when a nuclear material was last purified. However, a nuclear fuel is not purified after irradiation. New chronometry techniques need to be developed.

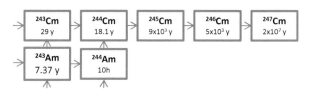

Fig. 1: *Neutron activation chart.*

The Cm isotopes ^{245}Cm and ^{246}Cm can only be formed by the neutron activation of ^{244}Cm (Fig. 1), even if the neutron flux in the reactor vary. Therefore, a mathematical relation exists between $^{245}Cm/^{246}Cm$ and $^{244}Cm/^{246}Cm$ ratios. To demonstrate that relationship irradiated nuclear fuel samples were analyzed. The samples were dissolved and Cm was purified using extraction chromatographic resins. The Cm isotopes were then measured by AMS at ETH Zurich. The measured relationship between $^{245}Cm/^{246}Cm$ and $^{244}Cm/^{246}Cm$ ratios is shown for samples of known age in Fig. 2. The ratio $^{245}Cm/^{246}Cm$ is almost constant for hundreds of years. Therefore, the $^{245}Cm/^{246}Cm$ ratio (Fig. 2)

can be used to estimate the $^{244}Cm/^{246}Cm$ at origin (R_0) for unknown samples.

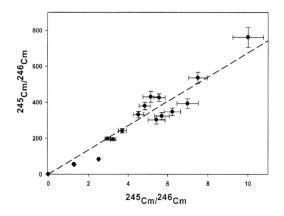

Fig. 1: $^{245}Cm/^{246}Cm$ as a function of $^{244}Cm/^{246}Cm$ ratios in nuclear fuels.

The actual $^{244}Cm/^{246}Cm$ ratio (R_1) is determined from the sample. With R_0 and R_1 values, the decay time (t) of the fuel can be simply calculated using the Bateman equation ($\lambda = \ln(2)/T_{1/2}$ with $T_{1/2}$ of ^{244}Cm = 18.11 yr^{-1}):

$$t = \frac{-ln\left(\frac{R_1}{R_0}\right)}{\lambda}$$

The main advantages of this method are that no Cm tracer is required, a single element is measured, it can be applied to unpurified irradiated fuels, and there is no issue of differing chemistry in a radiochronometer pair. An assumption of the technique is that there is minimal Cm present in the fuel prior to irradiation. The main drawback of this radiochronometer is the precision (currently ± 7 yr). Adding more data to the model will improve the precision.

---

[1] *Canadian Nuclear Laboratories, Chalk River, Canada*

MATERIAL SCIENCE

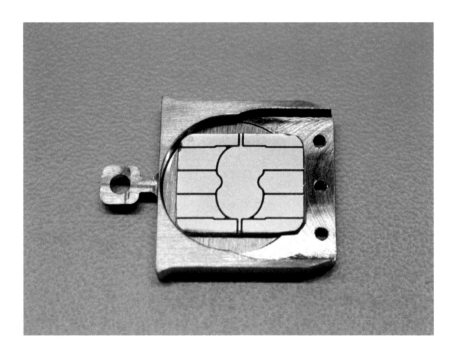

Time resolved MeV-SIMS yield measurements

Energy scaling of MeV-SIMS yields

Analysis of oxynitride thin films grown by PCLA

High power impulse magnetron sputtering

Lead sulfide colloidal quantum dot solids

Analysis of III-V quaternary semiconductors

Nuclear reaction profiling in heavy compounds

Composition of high performance rubber

TIME RESOLVED MEV-SIMS YIELD MEASUREMENTS

Performance tests of CHIMP with biochemical key substances

K.-U. Miltenberger, M. Döbeli, H.-A. Synal

MeV-SIMS is an imaging technique for the distribution of large molecules on a surface. Test measurements were performed on three standard organic substances: Arginine (amino-acid), Leu-enkephalin (peptide) and PEG (polyethylene glycol).

Samples were measured using the CHIMP MeV-SIMS setup [1,2] with 28 MeV $^{197}Au^{7+}$ primary ions. The capillary microbeam was scanned across a fresh, square area of 200 x 200 μm^2 using a total fluence of about 10^{10} ions/cm^2 with a primary ion rate of approximately 1 kHz which was periodically checked in the gas ionization detector. The total measurement time was 4000 s per sample. Accumulated spectra are shown in Fig. 1.

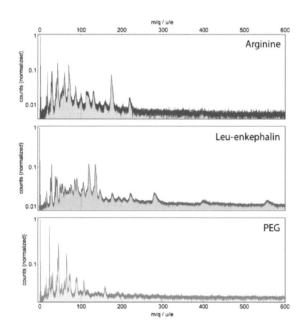

Fig. 1: *Accumulated mass spectra obtained for three organic samples .*

From the recorded spectra and primary ion fluences, secondary ion yields for selected peaks in the mass spectra were calculated and are listed in Tab. 1.

Sample	Fluence / ions/cm²	Mass / u	Yield
Arg.	1.02 ± 0.03	175	0.015 ± 0.002
L.-E.	1.07 ± 0.03	136	0.065 ± 0.003
PEG	1.00 ± 0.03	64	0.038 ± 0.005

Tab. 1: *Secondary Ion Yields for selected peaks in the mass spectra of the three samples.*

Spectra were acquired in single-event list mode. In order to observe also the development of yields with increasing fluence and to gain information on the destruction cross-section of the primary particles, data was rebinned into 100 s intervals. As shown in Fig. 2, the yields are fairly constant with increasing primary fluence and despite the sample being damaged no significant decrease in the signal is observed. Obviously, higher fluences are needed to obtain information on destruction radii of primary ions.

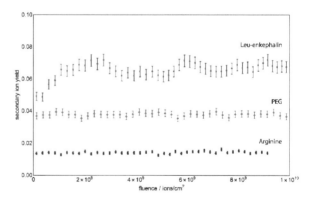

Fig. 2: *Development of secondary ion yield for the three organic samples with increasing primary ion fluence.*

[1] M. Schulte-Borchers et al., NIMB 380 (2016) 94

[2] K.-U. Miltenberger et al., NIMB 412 (2017) 185

ENERGY SCALING OF MEV-SIMS YIELDS

Comparing secondary ion yields for a range of primary ion energies

N. Brehm, K.-U. Miltenberger, M. Döbeli, H.-A. Synal

In secondary ion mass spectrometry (SIMS) a primary ion beam is used to induce the emission of secondary particles from a sample surface. To study the scaling of secondary ion yields with primary ions in the MeV energy range measurements were performed on the CHIMP MeV-SIMS setup [1].

To assess the measurement reproducibility in a first step, the total secondary ion yield of a single polyethylene glycol (PEG) sample with a 15 MeV iodine primary ion beam was measured on five different days. Fig. 1 shows the resulting mass spectra. They were normalized to the area of the peak at 107 u/e.

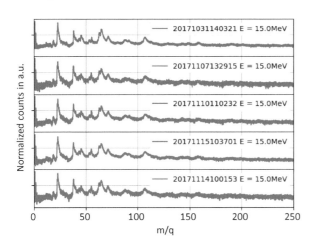

Fig. 1: *Secondary ion mass spectra of PEG with a 15 MeV iodine primary ion beam.*

In Fig. 2 the total yields are shown. A mean yield of 3.52 secondary ions per primary ion was determined with a standard deviation of 0.16. This corresponds well with the relative error of 5% assumed for the primary ion count rate measurement. Overall, a very satisfactory repeatability of the setup is proven.

A set of yield measurements of PEG was then done for different primary iodine ion energies between 15 and 38 MeV. In Fig. 3 the yield as a function of the electronic energy loss of the primary ion is presented. As expected, a clear trend of increasing yield with electronic energy deposition in the organic material is observed. In a next step, yield measurements with MeV cluster ion beams will be conducted.

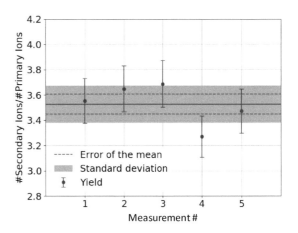

Fig. 2: *Yield measurements with the same sample and primary ion beam on different days.*

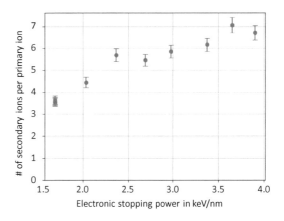

Fig. 3: *Total secondary ion yields of a PEG sample versus electronic energy loss of the ^{127}I primary ion beam.*

[1]   M. Schulte-Borchers et al., NIM B 380 (2016) 94

ANALYSIS OF OXYNITRIDE THIN FILMS GROWN BY PRCLA

Catalysts for photoelectrochemical water splitting

F. Haydous[1], D. Pergolesi[1], T. Lippert[1], M. Döbeli

Photocatalytic solar water splitting is of considerable interest as it appears to be a promising solution for the energy crisis. In this work, oxynitride photocatalyst thin films are deposited by Pulsed Reactive Crossed-beam Laser Ablation (PRCLA, Fig. 1).

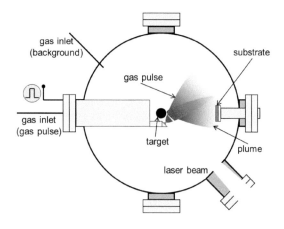

Fig. 1: *Schematics of the PRCLA technique.*

Determining the composition of the thin films and their N to O ratio is essential in order to taylor their photoelectrochemical activity.

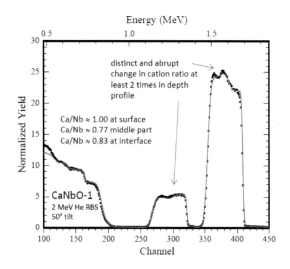

Fig. 2: *^{4}He RBS spectrum of a CaNbO$_2$N film.*

2 MeV 4He RBS analysis with a tilt angle showed that for CaNbO$_2$N thin films deposited on MgO substrate using a Ca$_2$Nb$_2$O$_7$ target, a variation in the Ca/Nb ratio can be found throughout the film (Fig. 2). This can be explained by non-congruent transfer of the cations to the substrate and subsequent diffusion. The use of a Ca enriched target could be a possible solution.

The combination of RBS and ERDA also allowed the accurate determination of the composition of epitaxial LaTiO$_2$N (LTON) films. In an attempt to increase the N content LTON films were post annealed in NH$_3$ at different temperatures. The photoactivity increased the most when annealed at 500°C for 2 hours (Fig. 3). By comparing the N/O ratio of this film with the untreated material, an increase is observed which explains the performance enhancement.

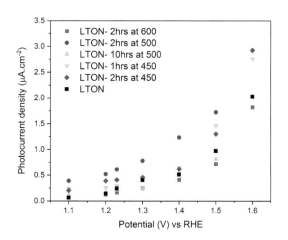

Fig. 3: *Photoelectrochemical performance of LTON films with/out post-annealing in NH$_3$.*

[1] *Paul Scherrer Institute, Villigen*

HIGH POWER IMPULSE MAGNETRON SPUTTERING

ERDA and RBS analysis of HiPIMS deposited carbon layers

R. Ganesan [1], K. Thorwarth[1], H.J. Hug[1], M. Döbeli

Tetrahedral amorphous carbon films with an sp3 content of >80% have been produced by high power impulse magnetron sputtering (HiPIMS) in a mode where the sputtering is mixed with an arc discharge (Fig. 1). Short-lived cathode spots form in the magnetic racetrack and produce large numbers of carbon ions. The spots move rapidly, inhibiting the formation of macro particles. Optimization of electrical input power, gas pressure, and deposition temperature are critical for obtaining carbon films with high sp3 content correlated with high film density.

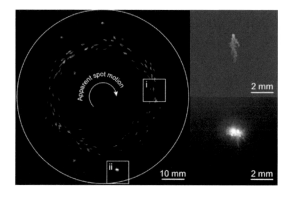

Fig. 1: *High current magnetron discharge with short-lived cathodic arc [1].*

To produce films with different sp3 content, electrical power to the sputtering carbon target and Ar gas pressure were varied at constant sputtering bias on the substrate. For tool coating, carbon films should be hydrogen-free to obtain high hardness (>35 GPa) and durability in harsh mechanical environments. After deposition the material composition was determined by 2 MeV ^{4}He RBS and ERD. Fig. 2 shows an example of an ERD spectrum from a batch of samples for which the hydrogen content in the films is limited to less than 0.75 at%. The Ar content of the films induced by the deposition process is between 3 and 4 at% as shown by the RBS measurements (Fig. 3). There is no evidence of any other impurity in

the films which was proven by heavy ion ERDA measurements not shown here.

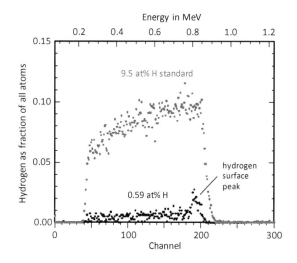

Fig. 2: *Hydrogen depth profile in an amorphous carbon film measured by 2 MeV He-ERD.*

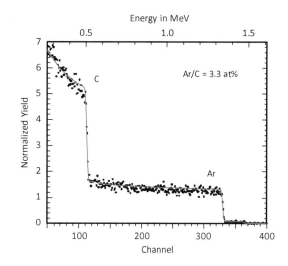

Fig. 3: *RBS spectrum of the same sample shown in Fig. 2.*

[1]   R. Ganesan et al., J. Phys. D48 (2015) 442001

[1] *Nanoscale Materials Science, EMPA Dübendorf*

LEAD SULFIDE COLLOIDAL QUANTUM DOT SOLIDS

Lead to sulfur ratios in photovoltaic materials

D.M. Balazs[1], K.I. Bijlsma[1], H.-H. Fang[1], D.N. Dirin[2], M.V. Kovalenko[2,3], M.A. Loi[1], M. Döbeli

Colloidal quantum dot (CQD) solids have great potential use in solar cells and photodetectors. Improvement of p-type layer conductivity in PbS CQD solids has been achieved by ligand exchange with chalcogenide salts, confirmed by transport properties of field-effect transistors (FETs) [1]. Excess sulfur is introduced as anhydrous sodium bisulfide (NaHS). A reference device (no sulfide added) shows the usual electron-dominated transport. Adding sulfur initially increases both electron and hole currents, then suppresses electron, but further increases hole conductivity (Fig. 1).

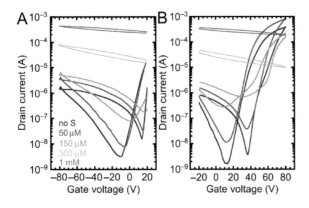

Fig. 1: *Transport properties measured in SiO_2-gated FETs. p-channel (A) and n-channel (B) transfer curves as a function of stoichiometry.*

To understand the effect of the treatment, film composition has been determined by RBS (Fig. 2). The stoichiometry as a function of NaHS level is shown in Fig. 3. The reference films show a large excess of lead and 14% iodine. With increasing sulfide level, the iodide content decreases and drops to trace amounts at 200 mM NaHS. The sulfur content shows an opposite trend, confirming that added sulfur predominantly replaces iodine at the CQD surface. The sample treated with 200 mM sulfide is close to stoichiometric. Further increase turns the material to sulfur-rich.

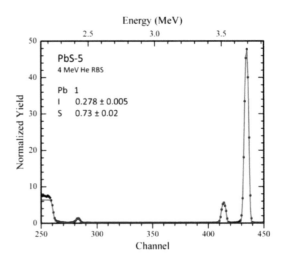

Fig. 2: *RBS spectrum for one of the samples.*

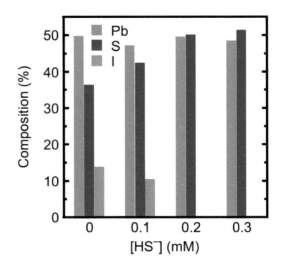

Fig. 3: *Film stoichiometry determined by RBS. A trend from lead-rich to sulfur-rich is observed as a function of the amount of NaHS[-].*

[1]   D. M. Balazs et al., Science Advances (2017) 3 aao1558

[1] *Advanced Materials, University of Groningen*
[2] *D-CHAB, ETHZ*
[3] *Empa Dübendorf*

ANALYSIS OF III-V QUATERNARY SEMICONDUCTORS

Determination of In and As concentrations in GaInAsSb

O. Ostinelli [1], W. Quan[1], A. K. M. Arabhavi[1], C. R. Bolognesi[1], M. Döbeli

The ternary III-V semiconductor alloy GaAsSb:C has been successfully used in DHBTs (double heterojunction bipolar transistors). However, InP/GaInAsSb DHBTs benefit from a higher DHBT cutoff frequency f_T [1] since the incorporation of In into the p-type GaAsSb:C increases the minority electron mobility.

In this project, the quaternary alloy $Ga_xIn_{1-x}As_ySb_{1-y}$:C was deposited on semi-insulating InP:Fe substrates by metal-organic vapour-phase epitaxy (MOVPE) in the centre for micro- and nanoscience FIRST (Fig. 1).

Fig. 1: *MOVPE Aix 200/4 in FIRST laboratory.*

Beside the GaInAsSb:C growth complexity, the determination of the alloy composition is complicated. In other quaternary alloys (e.g. AlGaAsSb), the composition can be obtained from the lattice constant via high resolution x-ray diffraction measurements and the material band gap measured by photoluminescence. For $Ga_xIn_{1-x}As_ySb_{1-y}$:C the described procedure does not work, since multiple combinations of In and As concentrations in the range of interest $x > 0.5$ and $y > 0.5$ lead to the same band gap and lattice constant (Fig. 2) [2]. With the enhanced mass resolution of 5 MeV He RBS it is possible to separate Ga from As and Sb from In (Fig. 3).

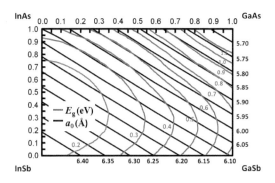

Fig. 2: *Contour lines of constant band gap (red lines) and constant lattice constant (black lines) for GaInAsSb.*

The In content in the film cannot be directly determined by RBS, as it also exists in the InP substrate. However, it can be deduced from the chemical valences in the compound.

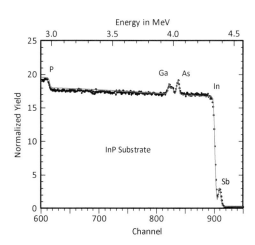

Fig. 3: 5 MeV RBS spectrum of a GaInAsSb film on InP. Ga, As, and Sb are separated.

[1] R. Flückiger et al., IEEE Electron Device Lett. 35 (2014) 166

[2] K. Shim et al., J. Appl. Phys. 88 (2000) 7157

[1] *Millimeter-Wave Electronics Group, ETHZ*

NUCLEAR REACTION PROFILING IN HEAVY COMPOUNDS

Oxygen analysis in coatings by the $^{16}O(d,\alpha)^{14}N$ exothermal reaction

M. Döbeli, J. Ast [1], M. Gindrat[2], A. Dommann[3], X. Maeder[1], A. Neels[4], J. Ramm[5], K. von Allmen[4]

The analysis of light elements in heavy matrices by MeV ion beam techniques is in general problematic due to the large scattering cross-section of charged particles by heavy elements, as they cause substantial unwanted background. In this project we studied the possibility to use the $^{16}O(d,\alpha)^{14}N$ nuclear reaction for depth profiling of oxygen [1] in superalloy oxide films. For this kind of material the accessible depth range of heavy ion ERDA is limited to 200-300 nm which calls for alternative analysis techniques. Two test samples an anodically grown 500 nm thick Ta_2O_5 film on Ta and a several micron thick superalloy oxide film on an alumina substrate, were used. Fig. 1 shows the α energy spectrum obtained from the Ta_2O_5 film with a 1 MeV deuterium beam. Simulations were performed with the SIMNRA software [2].

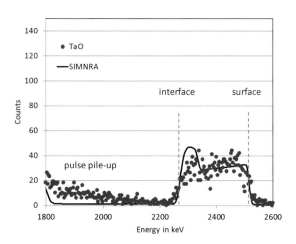

Fig. 1: α *energy spectrum obtained from a 500 nm Ta_2O_5 film on tantalum substrate.*

The shape and yield of the spectrum is quite well reproduced by the simulation. The main background is due to pulse pile-up from the orders of magnitude more intense elastic scattering of deuterons from Ta appearing in the low energy part of the spectrum. Therefore, the

data had to be acquired at very low α count rate. In Fig. 2 the α energy spectrum of the superalloy oxide film obtained under the same conditions as for the first Ta_2O_5 sample is displayed. Again the SIMNRA simulation reproduces the oxygen depth profile well down to a depth of about 1 µm where the background from pulse pile-up starts to overlap.

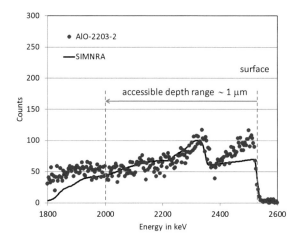

Fig. 2: α energy spectrum *of a superalloy oxide film on alumina.*

While the wiggle appearing at approx. 2350 keV is caused by the energy dependent reaction cross-section, the yield enhancement close to the surface seems to be due to a real feature of the oxygen depth profile of the film.

[1]   L. Wielunski, A. Barez, NIMB 111 (1973) 605

[2]   M. Mayer, Report IPP 9/113, Max-Planck-Institut, Garching (1997)

[1] *Empa Thun*

[2] *Oerlikon Metco AG, Wohlen*

[3] *Empa St. Gallen*

[4] *EMPA Dübendorf*

[5] *Oerlikon Surface Solutions AG, Balzers*

COMPOSITION OF HIGH PERFORMANCE RUBBER

Influence of sole composition on climbing shoe traction and adhesion

T.K. Jenny[1], M.Döbeli

This work was performed as part of a "Maturitätsarbeit" at the Kantonsschule Uster. The objective of the project was to investigate the properties of different light climbing shoes and a number of common training shoes to correlate their performance with material composition and make. As a contribution to this study the elemental composition of sole materials was determined by RBS and PIXE.

Fig. 1: *Rubber specimens are prepared and mounted on the RBS/PIXE sample holder.*

Soles of climbing shoes are made from rubber. The number of sulfur bridges in volcanized elastomers has a strong influence on their elasticity. The content of minor elements in 8 different sole samples was therefore measured. Figures 2 and 3 show RBS and PIXE spectra of one rubber material.

The sulfur content in all investigated materials was found to be between 0.1 % and 0.9 % which is lower than expected. This indicates that the high performance polymers are not mainly connected by sulfur bridges but rather by a second polymer group.

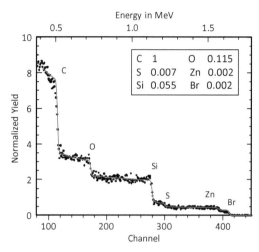

Fig. 2: *2 MeV He RBS spectrum of a high performance rubber elastomer.*

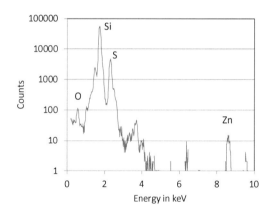

Fig. 3: *PIXE spectrum of the same material shown in Fig. 2.*

Final results of this project which mainly consists in the performance and correlation of mechanical material tests will be presented in the "Maturitätsarbeit" of Tobias Jenny.

---

[1] *Kantonsschule Uster*

EDUCATION

^{14}C and the protection of Cultural Heritage

National future day 2017 at LIP

Visiting archaeological excavations in Zurich

Automated reduction oven controller

^{14}C AND THE PROTECTION OF CULTURAL HERITAGE

First interdisciplinary conference

I. Hajdas, H.-A. Synal, E. Huysecom[1], A. Mayor[1], M.-A. Renold[2],

Radiocarbon dating is a technique used to measure the age of a material containing carbon. Following technical AMS developments over the last decades, the method now requires only small amounts of material, therefore, becoming attractive in the detection of forgeries. Private persons, conservators, antiquities dealers and auction houses often request a ^{14}C analysis of antique objects and radiocarbon laboratories might analyze such samples without inspecting the whole object. Sometimes, even if objects are brought to the laboratories their provenances remain unknown or even suspicious. In some cases this might be obvious from the appearance of the objects, for example their packing (Fig. 1) but others cannot be detected easily.

Huysecom et al. [1] addressed this issue and proposed a process that would support a unified approach of all radiocarbon laboratories to analysis of antiquities.

Fig. 1: *Objects submitted to the laboratory for ^{14}C analysis were packed in garbage bags.*

The conference, sponsored by SNF, took place at ETH on 16-17th November 2017 [2]. The presentations and discussions focused on issues related to the use of ^{14}C method to date objects, which are of value to cultural heritage (Fig. 2). Forty-five participants included archeologists,

radiocarbon researchers, museum curators, conservation scientists, lawyers, police officers, journalists and students from multidisciplinary fields.

Fig. 2: *Summary of presentation by E. Huysecom highlights the acute problem of looting.*

Final discussion resulted in setting a first draft of guidelines for accepting antique samples [2]. The next Radiocarbon conference will be an opportunity to continue discussions among the ^{14}C laboratories.

[1]   E. Huysecom et al., Radiocarbon. 59 (2017) 559

[2]   http://www.ams.ethz.ch/radiocarbon-dating-and-protection-of-cultural-heritage.html

---

[1] *Laboratory Archaeology and Population in Africa, University of Geneva*
[2] *Art-Law Center, University of Geneva*

NATIONAL FUTURE DAY 2017 AT LIP

Introducing our kids to the world of science

M. Christl, C. Vockenhuber, H.-A. Synal

The goal of National Future Day is to promote open career and life planning for school children regardless of gender. Therefore, on Future Day, hundreds of businesses, organizations, professional and academic schools across Switzerland invited girls and boys between 10 and 13 years to accompany a relative to work or to take part in an exciting special project [1].

Fig. 1: *Rector Sarah Springman opens the official Future Day at ETH Zurich [2].*

During the morning of November 9th 2017, ETH Zurich opened its doors for kids to experience the world of science, where their parents or relatives are working in. They could choose between several programs and so discover very different careers. The kids were very excited about their projects and Future Day at ETH Zurich was a great success.

In 2017, LIP did not officially offer a project for Future Day. However, as many LIP group members have kids in the right age their kids joined the official Future Day in the morning and after lunch some of them continued with their private and unofficial Future Day at LIP.

Together with their parents (real scientists!) they could explore the fascinating world of AMS physicists (Fig. 2).

Fig. 2: *Future Day continued at LIP.*

With the help of our kids we checked whether the charging system of the Tandy accelerator was working correctly after its revision (Fig. 3). They helped us reading and plotting the voltages and measured currents with the four multimeters connected to the charging system of the accelerator. It was a great (Future) day for them and for their parents!

Fig. 3: *Checking the charging system of the accelerator.*

[1] www.nationalerzukunftstag.ch/de/home/
[2] www.ethz.ch/de/news-und-veranstaltungen/veranstaltungen/zukunftstag.html

VISITING ARCHAEOLOGICAL EXCAVATIONS IN ZURICH

School project helps to date wooden stump

I. Hajdas, C. Lehmann[1], C. Thüring[1], B. Andres[2], P. Moser[2], D. Möckli[3], M. Maurer, M. Röttig

Zurich has a long history of human settlements, which combined with present urban activities provides excellent opportunities for archeologists. Also, the public profits from such developments because all excavations provide opportunities to visit. As a part of the Quaternary Dating Methods lecture [1], an excursion was organized in co-operation with Zurich Kantonsarchaeologie and Stadt Zurich.

Building of the new Children's Hospital in Zurich is underway near Burghölzli, a wooded hill in the district of Riesbach of southeastern Zürich. During the fall of 2017 prospection and archeological excavations by Kantonsarchaeologie found Bronze Age remains of fire places. The second part of an excursion was at another excavation near Burghölzli performed by archaeology of Stadt Zurich. A construction site close to the early Iron Age (800–450 BC) burial sites discovered in 1832 uncovered remains in the surrounding of a late Bronze Age settlement [2].

The wooden stump found at the site (Fig. 2) was sampled by students, prepared and analyzed as a part of a one-week project.

Fig. 2: *August-Forel-Strasse, Zurich. Tree stump found at the site was sampled during the student excursion.*

The age obtained for the outermost fragments of wood resulted in the ^{14}C age of 5 ka, i.e. to the time when the lake settlements flourished, also at Lake Zurich.

[1] 651-4901-00L Quaternary Dating Methods, ETHZ
http://www.vvz.ethz.ch/Vorlesungsverzeic hnis/lerneinheit.view?lerneinheitId=11582 5&semkez=2017W&lang=en

[2] https://www.stadt-zuerich.ch/hbd/de/index/archaeologie_de nkmalpflege_u_baugeschichte/stadtarchae ologie/fundstellen/August-Forel-Str_2017.html

Fig. 1: *Zurich, August-Forel-Strasse, 'remains' of wooden stumps in sediment. Sediment is darker due to organic input from the decayed wood.*

[1] *Kanotonsschule Olten*
[2] *Stadt Zurich, Amt für Städtebau, Archäologie*
[3] *Kantonsarcheologie Zurich*

AUTOMATED REDUCTION OVEN CONTROLLER

Controller unit for automatic reduction of EA copper tubes

R. Schlatter, L. Wacker

For radiocarbon analysis, organic samples are combusted in an elemental analyzer (EA) for conversion into CO_2, either for graphitization with AGE or the direct measurement with a gas interface system (GIS). Excess oxygen is removed in a so-called reduction tube via the oxidation of copper. These reduction tubes have to be regenerated after a certain amount of samples.

Here, we present a new reduction oven controller (ROC) that automatizes the regeneration of the copper filled reduction tube. The reduction tube is placed in a tube furnace and heated to a temperature of 300°C for a selectable time. At the same time forming gas (10% H_2 in N_2) is fed through the tube and the copper is reduced. Subsequently, the tube furnace cools down and an inert gas (N_2) is flushed through the system in order to avoid re-sorption of gases/aerosols by the copper.

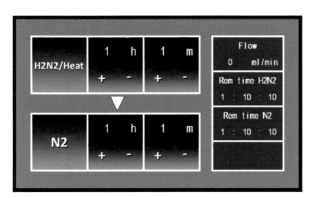

Fig. 1: *Touch screen user interface of the ROC: the time for copper reduction and subsequent flushing can be set on the left (hours and minutes). Information about the status of the system is given in a status display on the right.*

The ROC controls the temperature in the tube furnace and switches between the gases used for the reduction and the subsequent cooling process. The timing of the processes can be set on a touch screen (Fig. 1).

Fig. 2: *Photograph of the installed reduction oven setup with an inserted reduction tube (right) with the new controller in the middle and a reduction tube waiting for its regeneration (left).*

After the desired H_2 reduction and N_2 flushing times are set, the reduction at 300°C is started by pressing the H_2N_2/Heat-button. The reduction stops automatically and N_2 is flushed through the system while the oven temperature drops.

The installed setup at LIP is shown in Fig. 2. It has been in routine operation for several months and has proved to be a success. The main advantages involved with the automation are a higher efficiency with regard to the number of regenerated reduction tubes and that gas is no longer wasted.

PUBLICATIONS

N. Akçar, S. Ivy-Ochs, V. Alfimov, F. Schlunegger, A. Claude, R. Reber, M. Christl, C. Vockenhuber, A. Dehnert, M. Rahn and C. Schlüchter
Isochron-burial dating of glaciofluvial deposits: first results from the Swiss Alps
Earth Surface Processes and Landforms (2017)

N. Akçar, V. Yavuz, S. Yeşilyurt, S. Ivy-Ochs, R. Reber, C. Bayrakdar, P. W. Kubik, C. Zahno, F. Schlunegger and C. Schlüchter
Synchronous last glacial maximum across the Anatolian peninsula
Geological Society, London, Special Publications **433** (2017) 251 - 269

J. Ast, M. Döbeli, A. Dommann, M. Gindrat, X. Maeder, A. Neels, P. Polcik, M. Polyakov, H. Rudigier, K. von Allmen, B. Widrig and J. Ramm
Synthesis and characterization of superalloy coatings by cathodic arc evaporation
Surface and Coatings Technology **327** (2017) 139 - 145

D. M. Balazs, K. I. Bijlsma, H.-H. Fang, D. N. Dirin, M. Döbeli, M. V. Kovalenko and M. A. Loi
Stoichiometric control of the density of states in PbS colloidal quantum dot solids
Science advances **3** (2017) 1 - 7

R. Bao, M. Strasser, A. McNichol, N. Haghipour, C. McIntyre, G. Wefer and T. Eglinton
Rolling in the Deep: Tectonically-triggered sediment and carbon export to the Hadal zone
EGU General Assembly Conference Abstracts **19** (2017) 3718

C. Baroni, S. Casale, M. C. Salvatore, S. Ivy-Ochs, M. Christl, L. Carturan, R. Seppi and A. Carton
Double response of glaciers in the Upper Peio Valley (Rhaetian Alps, Italy) to the Younger Dryas climatic deterioration
Boreas **46** (2017) 783 - 798

M. Boxleitner, A. Musso, J. Waroszewski, M. Malkiewicz, M. Maisch, D. Dahms, D. Brandová, M. Christl, R. de Castro Portes and M. Egli
Late Pleistocene–Holocene surface processes and landscape evolution in the central Swiss Alps
Geomorphology **295** (2017) 306 - 322

G. Büttner, S. Populoh, W. Xie, M. Trottmann, J. Hertrampf, M. Döbeli, L. Karvonen, S. Yoon, P. Thiel, R. Niewa and A. Weidenkaff
Thermoelectric properties of $[Ca_2CoO_{3-\delta}][CoO_2]_{1,62}$ as a function of Co/Ca defects and Co_3O_4 inclusions
Journal of Applied Physics **121** (2017)

C. Cancellieri, F. Evangelisti, T. Geldmacher, V. Araullo-Peters, N. Ott, M. Chiodi, M. Döbeli and P. Schmutz
The role of Si incorporation on the anodic growth of barrier-type Al oxide
Materials Science & Engineering B **226** (2017) 120 - 131

S. Canulescu, M. Döbeli, X. Yao, T. Lippert, S. Amoruso and J. Schou
Nonstoichiometric transfer during laser ablation of metal alloys
Physical Review Materials **1** (2017) 073402/073401 - 073402/073407

N. Casacuberta, M. Christl, K. O. Buesseler, Y. Lau, C. Vockenhuber, M. Castrillejo, H.-A. Synal and P. Masqué
Potential releases of ^{129}I, ^{236}U, and Pu isotopes from the Fukushima Dai-ichi Nuclear Power Plants to the ocean from 2013 to 2015
Environmental Science & Technology **51** (2017) 9826 - 9835

M. Castrillejo, N. Casacuberta, M. Christl, J. Garcia-Orellana, C. Vockenhuber, H.-A. Synal and P. Masqué
Anthropogenic ^{236}U and ^{129}I in the Mediterranean Sea: First comprehensive distribution and constrain of their sources Science of the
Total Environment **593** (2017) 745 - 759

J. Chen, H. Chen, F. Hao, X. Ke, N. Chen, T. Yajima, Y. Jiang, X. Shi, K. Zhou, M. Döbeli, T. Zhang, B. Ge, H. Dong, H. Zeng, W. Wu and L. Chen
Ultrahigh Thermoelectric Performance in SrNb0. 2Ti0. 8O_3 Oxide Films at a Submicrometer-scale Thickness ACS
Energy Letters **2** (2017) 915 - 921

Y. Chen, M. Döbeli, E. Pomjakushina, Y. Gan, N. Pryds and T. Lippert
Scavenging of oxygen vacancies at modulation-doped oxide interfaces: Evidence from oxygen isotope tracing
Physical Review Materials **1** (2017) 052002/052001 - 052002/052006

M. Christl, N. Casacuberta, J. Lachner, J. r. Herrmann and H.-A. Synal
Anthropogenic ^{236}U in the North Sea–A Closer Look into a Source Region
Environmental Science & Technology **51** (2017) 12146 - 12153

A. Claude, N. Akçar, S. Ivy-Ochs, F. Schlunegger, P. Rentzel, C. Pümpin, D. Tikhomirov, P. W. Kubik, C. Vockenhuber, A. Dehnert, M. Rahn and C. Schlüchter
Chronology of Quaternary terrace deposits at the locality Hohle Gasse (Pratteln, NW Switzerland)
Swiss Journal of Geosciences **110** (2017) 793 - 809

A. Claude, N. Akçar, S. Ivy-Ochs, F. Schlunegger, P. K. W, A. Dehnert, J. Kuhlemann, M. Rahn and C. Schlüchter
Timing of early Quaternary gravel accumulation in the Swiss Alpine Foreland
Geomorphology **276** (2017) 71 - 85

B. Courel, P. Schaeffer, P. Adam, E. Motsch, Q. Ebert, E. Moser, C. Féliu, S. M. Bernasconi, I. Hajdas, D. Ertlen and D. Schwartz
Molecular, isotopic and radiocarbon evidence for broomcorn millet cropping in Northeast France since the Bronze Age
Organic Geochemistry **110** (2017) 13 - 24

R. Cusnir, M. Christl, P. Steinmann, F. Bochud and P. Froidevaux
Evidence of plutonium bioavailability in pristine freshwaters of a karst system of the Swiss Jura Mountains
Geochimica Et Cosmochimica Acta **206** (2017) 30 - 39

W. Dai, L.-T. Lee, A. Schütz, B. Zelenay, Z. Zheng, A. Borgschulte, M. Döbeli, W. Abuillan, O. V. Konovalov, M. Tanaka and D. A. Schlüter
Three-Legged 2, 2'-Bipyridine Monomer at the Air/Water Interface: Monolayer Structure and Reactions with Ni (II) Ions from the Subphase
Langmuir **33** (2017) 1646 - 1654

M. Dühnforth, A. L. Densmore, S. Ivy-Ochs, P. Allen and P. W. Kubik
Early to Late Pleistocene history of debris-flow fan evolution in western Death Valley using cosmogenic ^{10}Be and ^{26}Al
Geomorphology **281** (2017) 53 - 65

M. G. Fellin, C.-Y. Chen, S. D. Willett, M. Christl and Y.-G. Chen
Erosion rates across space and timescales from a multi-proxy study of rivers of eastern Taiwan
Global and Planetary Change **157** (2017) 174 - 193

X. Feng, J. E. Vonk, C. Griffin, N. Zimov, D. B. Montluçon, L. Wacker and T. I. Eglinton
14C Variation of Dissolved Lignin in Arctic River Systems
ACS Earth and Space Chemistry **1** (2017) 334 - 344

A. Fontana, G. Vinci, G. Tasca, P. Mozzi, M. Vacchi, G. Bivi, S. Salvador, S. Rossato, F. Antonioli, A. Asioli, M. Bresolin, F. Di Mario and I. Hajdas
Lagoonal settlements and relative sea level during Bronze Age in Northern Adriatic: Geoarchaeological evidence and paleogeographic constraints
Quaternary International **439** (2017) 17 - 36

H. Galinski, G. Favraud, H. Dong, J. S. T. Gongora, G. Favaro, M. Döbeli, R. Spolenak, A. Fratalocchi and F. Capasso
Scalable, ultra-resistant structural colors based on network metamaterials
Light: Science & Applications **6** (2017)

H. Galinski, A. Fratalocchi, M. Döbeli and F. Capasso
Light Manipulation in Metallic Nanowire Networks with Functional Connectivity
Advanced Optical Materials **5** (2017) 1 - 6

B. Giaccio, I. Hajdas, R. Isaia, A. Deino and S. Nomade
High-precision ^{14}C and ^{40}Ar/^{39}Ar dating of the Campanian Ignimbrite (Y-5) reconciles the time-scales of climatic-cultural processes at 40 ka
Scientific Reports **7** (2017) 1 - 10

L. M. Grämiger, J. R. Moore, V. S. Gischig, S. Ivy-Ochs and S. Loew
Beyond debuttressing: Mechanics of paraglacial rock slope damage during repeat glacial cycles
Journal of Geophysical Research: Earth Surface **122** (2017) 1004 - 1036

R. Grischott, F. Kober, M. Lupker, K. Hippe, S. Ivy-Ochs, I. Hajdas, B. Salcher and M. Christl
Constant denudation rates in a high alpine catchment for the last 6 kyrs
Earth Surface Processes and Landforms **42** (2017) 1065 - 1077

R. Grischott, F. Kober, M. Lupker, J. M. Reitner, R. Drescher-Schneider, I. Hajdas, M. Christl and
S. D. Willett
Millennial scale variability of denudation rates for the last 15 kyr inferred from the detrital ^{10}Be record of Lake Stappitz in the Hohe Tauern massif, Austrian Alps
The Holocene (2017) 1 - 14

K. Gückel, T. Shinonaga, M. Christl and J. Tschiersch
Scavenged ^{239}Pu, ^{240}Pu, and ^{241}Am from snowfalls in the atmosphere settling on Mt. Zugspitze in 2014, 2015 and 2016
Scientific Reports (2017) 1 - 9

C. Guerra-Nuñez, M. Döbeli, J. Michler and I. Utke
Reaction and Growth Mechanisms in Al_2O_3 deposited via Atomic Layer Deposition: Elucidating the Hydrogen Source
Chemistry of Materials **29** (2017) 8690 - 8703

I. Hajdas, L. Hendriks, A. Fontana and G. Monegato
Evaluation of preparation methods in radiocarbon dating of old wood
Radiocarbon **59** (2017) 727 - 737

U. Hanke, L. Wacker, N. Haghipour, M. Schmidt, T. Eglinton and C. McIntyre
Comprehensive radiocarbon analysis of benzene polycarboxylic acids (BPCAs) derived from pyrogenic carbon in environmental samples
Radiocarbon **59** (2017) 1103 - 1116

U. M. Hanke, C. M. Reddy, A. L. Braun, A. I. Coppola, N. Haghipour, C. P. McIntyre, L. Wacker, L. Xu,
A. P. McNichol, S. Abiven, M. W. Schmidt and T. I. Eglinton
What on Earth have we been burning? Deciphering sedimentary records of pyrogenic carbon
Environmental Science & Technology **51** (2017) 12972 - 12980

A. S. Hein, A. Cogez, C. M. Darvill, M. Mendelova, M. R. Kaplan, F. Herman, T. J. Dunai, K. Norton, S. Xu,
M. Christl and A. Rodés
Regional mid-Pleistocene glaciation in central Patagonia
Quaternary Science Reviews **164** (2017) 77 - 94

C. Heineke, R. Hetzel, C. Akal and M. Christl
Constraints on Water Reservoir Lifetimes From Catchment-Wide ^{10}Be Erosion Rates—A Case Study From Western Turkey
Water Resources Research (2017)

L. Hellmann, W. Tegel, J. Geyer, A. V. Kirdyanov, A. N. Nikolaev, Ó. Eggertsson, J. Altman, F. Reinig,
S. Morganti, L. Wacker and U. Büntgen
Dendro-provenancing of Arctic driftwood
Quaternary Science Reviews **162** (2017) 1 - 11

L. Hendriks, I. Hajdas, E. S. Ferreira, N. C. Scherrer, S. Zumbühl, M. Küffner, L. Wacker, H.-A. Synal and
D. Günther
Combined 14 C Analysis of Canvas and Organic Binder for Dating a Painting
Radiocarbon **60** (2017) 207 - 218

J. D. Hemingway, E. Schefuß, R. G. Spencer, B. J. Dinga, T. I. Eglinton, C. McIntyre and V. V. Galy
Hydrologic controls on seasonal and inter-annual variability of Congo River particulate organic matter source and reservoir age
Chemical Geology **466** (2017) 454 - 465

R. L. Hermanns, M. Schleier, M. Böhme, L. H. Blikra, J. Gosse, S. Ivy-Ochs and P. Hilger
Rock-Avalanche Activity in W and S Norway Peaks After the Retreat of the Scandinavian Ice Sheet
Workshop on World Landslide Forum (2017) 331 - 338

K. Hippe
Constraining processes of landscape change with combined in situ cosmogenic ^{14}C-^{10}Be analysis
Quaternary Science Reviews **173** (2017) 1 - 19

S. Ivy-Ochs, S. Martin, P. Campedel, K. Hippe, V. Alfimov, C. Vockenhuber, E. Andreotti, G. Carugati, D. Pasqual, M. Rigo and A. Vigano
Geomorphology and age of the Marocche di Dro rock avalanches (Trentino, Italy)
Quaternary Science Reviews **169** (2017) 188 - 205

S. Ivy-Ochs, S. Martin, P. Campedel, K. Hippe, C. Vockenhuber, G. Carugati, M. Rigo, D. Pasqual and A. Viganò
Geomorphology and Age of Large Rock Avalanches in Trentino (Italy): Castelpietra
Workshop on 4th World Landslide Forum (2017) 347 - 353

S. Jenatsch, J. Groenewold, M. Döbeli, R. Hany, R. Crockett, J. Lübben, F. Nüesch and J. Heier
Unexpected Equilibrium Ionic Distribution in Cyanine/C60 Heterojunctions
Advanced Materials Interfaces **4** (2017) 1 - 8

R. Jones, K. Norton, A. Mackintosh, J. Anderson, P. Kubik, C. Vockenhuber, H. Wittmann, D. Fink, G. Wilson, N. Golledge and R. Mckay
Cosmogenic nuclides constrain surface fluctuations of an East Antarctic outlet glacier since the Pliocene
Earth and Planetary Science Letters **480** (2017) 75 - 86

G. Jouvet, J. Seguinot, S. Ivy-Ochs and M. Funk
Modelling the diversion of erratic boulders by the Valais Glacier during the last glacial maximum
Journal of Glaciology **63** (2017) 487 - 498

E. Laloy, K. Beerten, V. Vanacker, M. Christl, B. Rogiers and L. Wouters
Bayesian inversion of a CRN depth profile to infer Quaternary erosion of the northwestern Campine Plateau (NE Belgium)
Earth Surface Dynamics **5** (2017) 331 - 345

M. Lavrieux, C. J. Schubert, T. Hofstetter, T. I. Eglinton, I. Hajdas, L. Wacker and N. Dubois
From medieval land clearing to industrial development: 800 years of human-impact history in the Joux Valley (Swiss Jura)
The Holocene (2017) 1 - 12

J. Lippold, M. Gutjahr, P. Blaser, E. Christner, M. L. de Carvalho Ferreira, S. Mulitza, M. Christl, F. Wombacher, E. Böhm, B. Antz, O. Cartapanis, H. Vogel and S. L. Jaccard
Deep water provenance and dynamics of the (de) glacial Atlantic meridional overturning circulation
Earth and Planetary Science Letters **458** (2017) 444 - 448

M. Lupker, J. Lavé, C. France-Lanord, M. Christl, D. Bourlès, J. Carcaillet, C. Maden, R. Wieler, M. Rahman, D. Bezbaruah and L. Xiaohan
^{10}Be systematics in the Tsangpo-Brahmaputra catchment: the cosmogenic nuclide legacy of the eastern Himalayan syntaxis
Earth Surface Dynamics **5** (2017) 429 - 449

K.-U. Miltenberger, A. M. Müller, M. Suter, H.-A. Synal and C. Vockenhuber
Accelerator mass spectrometry of ^{26}Al at 6MV using AlO– ions and a gas-filled magnet
Nuclear Instruments and Methods in Physics Research Section B: Beam Interactions with Materials and Atoms **406** (2017) 272 - 277

K.-U. Miltenberger, M. Schulte-Borchers, M. Döbeli, A. M. Müller, M. George and H.-A. Synal
MeV-SIMS capillary microprobe for molecular imaging Nuclear
Instruments and Methods in Physics Research Section B: Beam Interactions with Materials and Atoms **412** (2017) 185 - 189

G. Monegato, G. Scardia, I. Hajdas, F. Rizzini and A. Piccin
The Alpine LGM in the boreal ice-sheets game
Scientific Reports **7** (2017) 1 - 8

A. P. Moran, S. Ivy Ochs, M. Christl and H. Kerschner
Exposure dating of a pronounced glacier advance at the onset of the late-Holocene in the central Tyrolean Alps
The Holocene **27(9)** (2017) 1350 - 1358

A. M. Müller, M. Döbeli and H.-A. Synal
High resolution gas ionization chamber in proportional mode for low energy applications
Nuclear Instruments and Methods in Physics Research Section B: Beam Interactions with Materials and Atoms **407** (2017) 40 - 46

R. M. Newnham, B. V. Alloway, K. A. Holt, K. Butler, A. B. H. Rees, J. M. Wilmshurst, G. Dunbar and I. Hajdas
Last Glacial pollen–climate reconstructions from Northland, New Zealand
Journal of Quaternary Science (2017) 2 - 19

A. Ojeda-GP, C. W. Schneider, M. Döbeli, T. Lippert and A. Wokaun
Plasma plume dynamics, rebound, and recoating of the ablation target in pulsed laser deposition
Journal of Applied Physics **121** (2017) 1 - 13

M. Ostermann, S. Ivy-Ochs, D. Sanders and C. Prager
Multi-method (^{14}C, ^{36}Cl, ^{234}U/^{230}Th) age bracketing of the Tschirgant rock avalanche (Eastern Alps): implications for absolute dating of catastrophic mass-wasting
Earth Surface Processes and Landforms **42** (2017) 1110 - 1118

E. Perret, K. Sen, J. Khmaladze, B. P. P. Mallett, M. Yazdi-Rizi, P. Marsik, S. Das, I. Marozau, M. A. Uribe-Laverde, R. de Andrés Prada, J. Strempfer, M. Döbeli, N. Biskup, M. Varela, Y.-L. Mathis and C. Bernhard
Structural, magnetic and electronic properties of pulsed-laser-deposition grown SrFe$_{O–\delta}$ thin films and SrFeO$_{3–\delta}$/La$_2$/$_3$Ca$_1$/$_3$MnO$_3$ multilayers
Journal of Physics: Condensed Matter **29** (2017)

M. Pichler, W. Si, F. Haydous, H. Téllez, J. Druce, E. Fabbri, M. E. Kazzi, M. Döbeli, S. Ninova, U. Aschauer, A. Wokaun, D. Pergolesi and T. Lippert
LaTiOxNy Thin Film Model Systems for Photocatalytic Water Splitting: Physicochemical Evolution of the Solid–Liquid Interface and the Role of the Crystallographic Orientation
Advanced Functional Materials (2017) 1 - 18

M. Pichler, J. Szlachetko, I. E. Castelli, N. Marzari, M. Döbeli, A. Wokaun, D. Pergolesi and T. Lippert
Determination of Conduction and Valence Band Electronic Structure of LaTiOxNy Thin Film
ChemSusChem **10** (2017) 2099 - 2106

L.-B. Qian, P.-F. Li, B. Jin, D.-K. Jin, G.-Y. Song, Q. Zhang, L. Wei, B. Niu, C.-L. Wan, C.-L. Zhou, M. Arnold Milenko, D. Max, Z.-Y. Song, Z.-H. Yang, S. Reinhold, H.-Q. Zhang and X.-M. Chen
Transmission of electrons through the conical glass capillary with the grounded conducting outer surface
Wuli xuebao **66** (2017) 124101/124101 - 124101/124109

G. M. Raisbeck, A. Cauquoin, J. Jouzel, A. Landais, J.-R. Petit, V. Y. Lipenkov, J. Beer, H.-A. Synal, H. Oerter, S. J. Johnsen, J. P. Steffensen, A. Svensson and F. Yiou
An improved north-south synchronization of ice core records around the 41kyr ^{10}Be peak
Climate of the Past **13** (2017) 217 - 229

R. Reber, R. Delunel, F. Schlunegger, C. Litty, A. Madella, N. Akçar and M. Christl
Environmental controls on ^{10}Be-based catchment-averaged denudation rates along the western margin of the Peruvian Andes
Terra Nova (2017) 282 - 293

A. Romundset, N. Akçar, O. Fredin, D. Tikhomirov, R. Reber, C. Vockenhuber, M. Christl and C. Schlüchter
Lateglacial retreat chronology of the Scandinavian Ice Sheet in Finnmark, northern Norway, reconstructed from surface exposure dating of major end moraines
Quaternary Science Reviews **177** (2017) 130 - 144

S. Schneider, S. Bister, M. Christl, M. Hori, K. Shozugawa, H.-A. Synal, G. Steinhauser and C. Walther
Radionuclide pollution inside the Fukushima Daiichi exclusion zone, part 2: Forensic search for the "Forgotten" contaminants Uranium-236 and plutonium
Applied Geochemistry **85** (2017) 194 - 200

J. Schoonejans, V. Vanacker, S. Opfergelt and M. Christl
Long-term soil erosion derived from in-situ ^{10}Be and inventories of meteoric 10Be in deeply weathered soils in southern Brazil
Chemical Geology **466** (2017) 380 - 388

N. P. Stadie, E. Billeter, L. Piveteau, K. V. Kravchyk, M. Döbeli and M. V. Kovalenko
Direct Synthesis of Bulk Boron-Doped Graphitic Carbon
Chemistry of Materials **29** (2017) 3211 - 3218

A. Stubbins, P. J. Mann, L. Powers, T. B. Bittar, T. Dittmar, C. P. McIntyre, T. I. Eglinton, N. Zimov and R. G. Spencer
Low photolability of yedoma permafrost dissolved organic carbon
Journal of Geophysical Research: Biogeosciences **122** (2017) 200 - 211

Z. Talip, R. Dressler, J. C. David, C. Vockenhuber, E. Müller Gubler, A. Vögele, E. Strub, P. Vontobel and D. Schumann
Radiochemical Determination of Long-Lived Radionuclides in Proton-Irradiated Heavy-Metal Targets: Part I - Tantalum
Analytical Chemistry **89** (2017) 13541 - 13549

S. Tao, T. I. Eglinton, L. Zhang, Z. Yi, D. B. Montluçon, C. McIntyre, M. Yu and M. Zhao
Temporal variability in composition and fluxes of Yellow River particulate organic matter
Limnology and Oceanography (2017)

C. Terrizzano, E. G. Morabito, M. Christl, J. Likerman, J. Tobal, M. Yamin and R. Zech
Climatic and Tectonic forcing on alluvial fans in the Southern Central Andes
Quaternary Science Reviews **172** (2017) 131 - 141

C. Terrizzano, E. G. Morabito, M. Christl, J. Likerman, J. Tobal, M. Yamin and R. Zech
Climatic and Tectonic forcing on alluvial fans in the Southern Central Andes
Quaternary Science Reviews **172** (2017) 131 - 141

C. Vivo-Vilches, J. M. López-Gutiérrez, M. García-León, C. Vockenhuber and T. Walczyk
41 Ca measurements on the 1 MV AMS facility at the Centro Nacional de Aceleradores (CNA, Spain)
Nuclear Instruments and Methods in Physics Research Section B: Beam Interactions with Materials and Atoms **413** (2017) 13 - 18

C. Vockenhuber, K. Arstila, J. Jensen, J. Julin, H. Kettunen, M. Laitinen, M. Rossi, T. Sajavaara, M. Thöni and H. Whitlow
Energy loss and straggling of MeV Si ions in gases
Nuclear Instruments and Methods in Physics Research Section B: Beam Interactions with Materials and Atoms **391** (2017) 20 - 26

C. Wang, S. Hou, H. Pang, Y. Liu, H. W. Gäggeler, M. Christl and H.-A. Synal
^{239,240}Pu and ^{236}U records of an ice core from the eastern Tien Shan (Central Asia)
Journal of Glaciology **63** (2017) 929 - 935

C. Welte, L. Wacker, B. Hattendorf, M. Christl, J. Koch, C. Yeman, S. F. Breitenbach, H.-A. Synal and D. Günther
Optimizing the analyte introduction for ^{14}C laser ablation-AMS
The Royal Society of Chemistry **32** (2017) 1813-1819

C. Wirsig, S. Ivy-Ochs, J. M. Reitner, M. Christl, C. Vockenhuber, M. Bichler and M. Reindl
Subglacial abrasion rates at Goldbergkees, Hohe Tauern, Austria, determined from cosmogenic ^{10}Be and ^{36}Cl concentrations
Earth Surface Processes and Landforms **42** (2017) 1119 - 1131

H. Wittmann, F. Blanckenburg, M. Mohtadi, M. Christl and A. Bernhardt
The competition between coastal trace metal fluxes and oceanic mixing from the ^{10}Be/^{9}Be ratio: Implications for sedimentary records
Geophysical Research Letters (2017) 1 - 10

A. Wölfler, C. Glotzbach, C. Heineke, N.-P. Nilius, R. Hetzel, A. Hampel, C. Akal, I. Dunkl and M. Christl
Late Cenozoic cooling history of the central Menderes Massif: Timing of the Büyük Menderes detachment and the relative contribution of normal faulting and erosion to rock exhumation
Tectonophysics **717** (2017) 585 - 598

L. Wüthrich, C. Brändli, R. Braucher, H. Veit, N. Haghipour, C. Terrizzano, M. Christl, C. Gnägi and R. Zech
^{10}Be depth profiles in glacial sediments on the Swiss Plateau: deposition age, denudation and (pseudo-) inheritance
E&G Quaternary Science Journal **66** (2017) 57 - 68

B. Zollinger, C. Alewell, C. Kneisel, D. Brandová, M. Petrillo, M. Plötze, M. Christl and M. Egli
Soil formation and weathering in a permafrost environment of the Swiss Alps: a multi-parameter and non-steady-state approach
Earth Surface Processes and Landforms **42** (2017) 814 - 835

TALKS AND POSTERS

A. Andrews, C. Welte, L. Wacker, M. Christl, C. Yeman
Can red snapper live for half a century? Laser - ablation AMS reveals complete bomb ^{14}C signal in an otolith
Hungary, Debrecen, 10.07.2017, Radiocarbon in the Environment

B. Ausin, C. Magill, P. Wenk, G. Haug, C. McIntyre, N. Haghipour, D. Hodell, T.Eglinton
Hydrodynamic Influences on Multiproxy-based Paleoclimate Reconstructions from Marine Sediments
USA, New Orleans, 13.12.2017, AGU

B. Ausin, C. Magill, P. Wenk, G. Haug, C. McIntyre, N. Haghipour, D. Hodell, T. Eglinton
An additional wrinkle in the Elderfield proxy development curve
France, Paris, 13.08.2017, Goldschmidt

B. Ausin, C. Magill, P. Wenk, G. Haug, C. McIntyre, N. Haghipour, D. Hodell, L. Wacker, S. Bernasconi, T. Eglinton
The masked influence of hydrodynamic changes on paleoceanographic proxy records: implications for temporal control on paleoclimate interpretations
United Kingdom, Edinburg, 09.11.2017, Heriot-Watt University Seminar Series

B. Ausin, C. Magill, P. Wenk, G. Haug, C. McIntyre, N. Haghipour, D. Hodell, T. Eglinton
Age offsets between climate proxy signals in the Shackeleton Sites
Switzerland, Bern, 15.09.2017, 1st Bern Workshop on C analyses with MICADAS

R. Bao, M. Starsser, A.P.McNichole,N. Haghipour, C.McIntyre, G.Wefer, T.Eglinton
Rolling in the Deep: Tectonically-triggered sediment and carbon export to the Hadal zone
Italy, Florence, 17.09.2017, International Meeting of Organic Geochemistry

S. Binnie, T. Dunai, A. Dewald, S. Heinze, H. Wittmann, F. von Blanckenburg, R. Hetzel, M. Christl, M. Schaller, D. Fabel, S. Freeman, S. Xu, C. Spiegel, B. Bookhagen, S. Ivy-Ochs, K. Hippe, N. Akçar
A quartz reference material for terrestrial cosmogenic ^{10}Be and ^{26}Al measurements
Canada, Ottawa, 18.08.2017, AMS14

T.M. Blattmann, Z. Liu, K. Wen, S. Lin, J. Li, Y. Zhao, Y. Zhang, L. Wacker, N. Haghipour, M. Plötze, T.I. Eglinton
Geochemical tracing of sedimentary organic matter and associated clay minerals in the South China Sea
Switzerland, Davos, SGM 2017

M. Maisch, D. Brandová, M. Egli, S. Ivy-Och, M. Christl
The post-LGM deglaciation in Central and Southeast Switzerland: New insights from surface exposure dating
Austria, Vienna, 23.-28.04.2017, EGU General Assembly

N. Casacuberta, M. Christl, C. Vockenhuber, A.-M. Wefing, P. Masqué, M. Castrillejo Iridoy, M.R. van der Loeff, S. Yang, N. Gruber
^{129}I/^{236}U and ^{236}U/^{238}U as a dual tracer for water mass circulation in the North Atlantic and Arctic Ocean
France, Paris, 13.08.2017, Goldschmidt 2017

N. Casacuberta, M. Christl, C. Vockenhuber, M. Castrillejo Iridoy, P. Masqué, M.R. van der Loef
Distribution and fate of ^{129}I and ^{236}U in the German GEOTRACES expedition to the Arctic Ocean in 2015
USA, Honolulu, 26.02.2017, ASLO 2017

N. Casacuberta, M. Christl, M.R. van der Loeff, P. Masqué, J. Herrmann, J. Lachner, G. Henderson, C. Walther, H.-A. Synal
^{236}U and ^{129}I as a new tool to study ocean circulation in the North Atlantic and Arctic Ocean
The Netherlands, Texel, 24.04.2017, Seminar

N. Casacuberta, M. Christl, M.R. van der Loeff, P. Masqué, H.A. Synal
Long-lived artificial radionuclides (^{236}U, ^{129}I and Pu) as tracers of ocean processes: from Fukushima to the Arctic Ocean
Germany, Bremerhaven, 07.09.2017, Seminar

N. Casacuberta, M. Christl, C. Vockenhuber, A.-M. Wefing, P. Masqué, M. Castrillejo Iridoy, M.R. van der Loeff, S. Yang, N. Gruber
Long-lived artificial radionuclides (^{129}I and ^{236}U) as tracers of ocean circulation in the Arctic and North Atlantic Ocean
Germany, Heidelberg, 30.11.2017, Seminar

N. Casacuberta, Bollhalder, S.; A.-M. Wefing, Synal, H.-A.
Simple CO_2 extraction from seawater and lake samples for ^{14}C analysis
Canada, Ottawa, 13.08.2017, AMS14

S. Casale, C. Baroni, M.C. Salvatore, S. Ivy-Ochs, M. Christl, L. Carturan, R. Seppi, A. Carton
In situ ^{10}Be exposure ages document a double glaciers response to Younger Dryas in the upper Peio Valley (Rhaetian Alps, Italy)
India, New Dehli, 06.-11.11.2017, International Conference On Geomorphology

M. Castrillejo Iridoy, J. Garcia-Orellana, T.C. Kenna, N. Casacuberta, M.Q. Fleisher, P. Masqué
Anthropogenic ^{137}Cs, ^{237}Np, ^{239}Pu and ^{240}Pu in the Mediterranean Sea
France, Paris, 13.08.2017, Goldschmidt 2017

E. Chamizo, M. Villa-Alfageme, M. López-Lora, M, N. Casacuberta, T.C. Kenna, P. Masqué, M. Christl
A first transect of ^{236}U at the Equatorial Pacific
Canada, Ottawa, 13.08.2017, AMS14

C.-Y. Chen, S. Willett, J.A. West, S. Dadson, N. Hovius, M. Christl, B.J.H. Shyu
The Spatial and Temporal Patterns of Erosion Rate in the Southern Central Range of Taiwan
Corea, Jeju Island, 04.-08.09.2017, ASQUA Conference 2017

S. Willett, A.J. West, S.J. Dadson, N. Hovius, M. Christl, J.B. Shyu
Meteoric ^{10}Be/^{9}Be ratios in marine sedimentary records: Deciphering the mixing between their marine and terrestrial sources and influence of costal trace metal fluxes
USA, New Orleans, 15.12.2017, AGU 2017

M. Christl, N. Casacuberta, C. Vockenhuber, J. Lachner, J. Herrmann, M.R van der Loeff, H.-A. Synal
The distribution of ^{236}U and ^{129}I in the North Sea and the Arctic Ocean – implication for the input of radionuclides from nuclear reprocessing facilities
Canada, Ottawa, 17.08.2017, AMS14

M. Christl, C. Vockenhuber, N. Casacuberta, S. Maxeiner, J. Thut, H.-A. Synal
Quasi simultaneous measurements of $^{236}U/^{238}U$ ratios on the compact ETH Zurich Tandy AMS system
Germany, Mainz, 08.03.2017, DPG-17 Conference

M. Christl, N. Casacuberta, C. Vockenhuber, H.-A. Synal
The actinide program at ETH Zurich an overview
Canada, Ottawa, 12.08.2017, AMS14 Actinides Workshop

A.I. Coppola, D.B. Wiedemeier, V. Galy, N. Haghipour, U. Hanke, G. Nascimento, M. Usman, T. Blattmann, M. Reisser, C. Freymond, M. Zhao, B. Voss, E. Schefuß, L. Wacker, B. Peucker-Ehrenbrink, S. Abiven, M.W.I. Schmidt, T. Eglinton
Global Scale Fluvial Export of Persistent Particulate Black Carbon
USA, New London, New Hamsphire, 21.07.2017, Gordon Marine Chemistry Conference

D. Dahms, M. Egli, D. Fabel, D. Brandova, R. Portes, M. Christl
New ^{10}Be Exposure Ages For Pleistocene Glacial Stratigraphy, Southern Wind River Range, Wyoming, USA
USA, Seattle, 04.-07.11.2017, GSA Annual Meeting

X. Dai, M. Christl, M. Luo, H.-A. Synal
Rapid sample preparation methods for determination of actinides in biological and environmental samples by AMS
Canada, Ottawa, 15.08.2017, AMS14

Z. Kazi, M. Christl
Measurement of minor actinides by compact AMS for radio bioassay and nuclear forensic applications
Canada, Ottawa, 12.08.2017, AMS14 Actinides Workshop

D. De Maria, H.-A. Synal, A. Müller, S. Maxeiner
Beam Profile Monitor, Device design and phase space measurements
Germany, Mainz, 07.03.2017, DPG-17 Conference

D. De Maria, S. Fahrni, L. Wacker, H.-A. Synal
Design of a new Gas Ion Source interface for biomedical ^{14}C analyses
Switzerland, Bern, 13.09.2017, 1st MICADAS Workshop Bern

C. Degueldre, J. Fahy, O. Kolosov, R. Wilbraham, M. Döbeli, N. Renevier, J. Ball, S. Ritter
Mechanical properties of stainless steel cladding after irradiation
United Kingdom, Lancaster, 05.09.2017, Nuclear Academic Meeting

B. Dittmann, R. Buompane, M. Christl, T. Dunai, C. Feuerstein, K. Fifield, M. Fröhlich, S. Heinze, F. Marzaioli, C. Münker, A. Petraglia, T. Reich, D. Schönenbach, E. Strub, C. Sirignano, H.-A. Synal, P. Thörle-Pospiech, N. Trautmann, A. Wallner
Lab Intercomparison for the Establishment of a New Multi-Isotope Plutonium Standard
Canada, Ottawa, 17.08.2017, AMS14

N. Dlamini, A. Mayor, I. Hajdas
Tracking Humans: A bio-archaeological approach to the history of pre-historic populations in the Dogon country, Republic of Mali
Denmark, Aarhus, 20.-23.06.2017, 2nd Radiocarbon And Diet Conference

L. Eggenschwiler, I. Hajdas, V. Picotti, P. Cherubini, M. Saurer, G. Battista Via, S. Marabini
Radiocarbon dating and Dendrochronology for Statigraphic Units near Tebano, Senio Northern Apennines – Time frame of Climatic Fluctuation at the onset of the Younger Dryas
Austria, Vienna, 23.-28.04.2017, EGU General Assembly

X. Dai, M. Christl, M. Totland, N. Guerin, S. Burrel, Z. Kazi
Cm and Am separation in spent nuclear fuel for Cm measurement by Accelerator Mass Spectrometry
Canada, Ottawa, 15.08.2017, AMS14

K. Grant, V. Galy, N. Haghipour, L. Wacker, T. Eglinton, L. Derry
Iron Loss Promotes SOC Turnover on a Hawaiian Soil Gradient
France, Paris, 16.08.2017, Goldschmidt 2017

R. Grischott, F. Kober, M. Lupker, J.M. Reitner, R. Drescher-Schneider, I. Hajdas, M. Christl
Millennial scale variability of denudation rates for the last 15 kyrs inferred from the detrital ^{10}Be record of lake Stappitz in the Hohe Tauern massif, Austrian Alps
Austria, Vienna 23.-28.2017, EGU General Assembly

R. Grischott, F. Kober, S. Ivy-Ochs, K. Hippe, M. Lupker, M. Christl, C. Vockenhuber, C. Maden
Determining the age of Swiss Deckenschotter with cosmogenic isochron burial dating
Switzerland, Davos, 18.11.2017, 15th SGM 2017

R. Grischott, F. Kober, M. Lupker, J.M. Reitner, R. Drescher-Schneider, I. Hajdas, M. Christl
Millennial scale variability of denudation rates for the last 15 kyrs inferred from the detrital ^{10}Be record of lake Stappitz in the Hohe Tauern massif, Austrian Alps
Austria, Salzburg, 03.11.2017, 6th Symposium for Research in Protected Areas

I. Hajdas
Die ^{14}C-Datierungs-Methode im Dienste des mittelalterlichen Kulturerbes
Switzerland, Bern, 13.01.2017, Kolloquium Uni Bern

I. Hajdas
Radiocarbon dating method--methodological aspects and applications
Poland, Lublin, 03.03.2017, University of Marie Curie-Sklodowska

I. Hajdas
Radiocarbon based chronologies of past and present climatic records—an overview
Hungary, Debrecen, 3.-7.7.2017, 2nd International Radiocarbon in the Environment Conference

I. Hajdas
Bomb peak radiocarbon a tracer and dating tool—an overview
Lituania, Vilnius, 29.5.-2.6.2017, ENVIRA2017

I. Hajdas
Radiocarbon dating method for research and protection of cultural heritage — a perspective
Germany, Köln, 27.11.2017, TH Köln Ringvorlesung

I. Hajdas, J. Heinemeier and participants of MODIS project
^{14}C dating of mortar and importance of samples selection—reflections from the first MOrtar Dating Inter-comparison Study (MODIS)
Canada, Ottawa, 14.-18.08.2017, AMS14

I. Hajdas, M. Maurer, M.B. Röttig, P. Ohnsorg, J. Trumm
^{14}C dating of mortar at ETH Zurich
Switzerland, Bern, 13.-15.09.2017, 1st MICADAS Workshop

I. Hajdas; L. Hendriks, M. Maurer, M.B. Röttig, C. Welte, L. Wacker, H.-A. Synal
Radiocarbon dating and research for Cultural Heritage
Switzerland, Affolterna. A., 05.10.2017, Analytikertreffen 2017

I. Hajdas
Bomb Peak ^{14}C – can it be useful in detection of forgeries?
Austria, Vienna, 13.-16.11.2017, IAEA 2017

I. Hajdas
Effect of anthropogenic activities on atmospheric ^{14}C content and radiocarbon chronologies of the future
Austria, Vienna, 23.-28.04.2017, EGU General Assembly

I. Hajdas, M. Maurer, M.B. Röttig
Radiocarbon dating: from the sample to age model, and issues of roots, old wood and bayesian models
France, Ardèche, 22.-24.9.2017, 20e Colloque et rencontre annuelle--Peuplement humain et paléoenvironnement en Afrique

I. Hajdas, M. Küffner, C. McIntyre, N. Scherrer, E. Ferreira, H.-A. Synal
Microscale radiocarbon dating of paintings
Germany, Mainz, 07.03.2017, DPG-17 Conference

I. Hajdas, E. Ferreira, N. Scherrer, S. Zumbühl, M. Küffner, H.-A. Synal, D. Günther
^{14}C dating of paintings
Switzerland, Bern, 28.03.2017, Seminar

I. Hajdas, E. Ferreira, N. Scherrer, S. Zumbühl, M. Küffner, H.-A. Synal, D. Günther
Radiocarbon dating of art objects
Germany, Köln, 19.06.2017, Seminar

I. Hajdas, E. Ferreira, N. Scherrer, S. Zumbühl, M. Küffner, L. Wacker, H.-A. Synal, D. Günther
A new approach to radiocarbon dating of paintings
Canada, Ottawa, 13.08.2017, AMS14

I. Hajdas, E. Ferreira, N. Scherrer, S. Zumbühl, M. Küffner, L. Wacker, H.-A. Synal, D. Günther
A new approach to radiocarbon dating of paintings
Switzerland, Bern, 13.09.2017, 1st MICADAS Workshop Bern

R. L. Hermanns, M. Schleier, M. Böhme, L.H. Blikra, J. Gosse, S. Ivy-Ochs, P. Hilger
Rock-avalanche activity in W and S Norway peaks after the retreat of the Scandinavian Ice Sheet
Slovenia, Ljubljana, 29.5.-2.6.2017, 4th World Landslide Forum

R. Hetzel, N.-P. Nilius, C. Glotzbach, A. Wölfler, A. Hampel, C. Akal, M. Christl
Spatial patterns of erosion and landscape evolution in the central Menderes Massif (Western Turkey) revealed by cosmogenic ^{10}Be
Austria, Vienna, 23.-28.04.2017, EGU General Assembly

R. Hetzel, N.-P. Nilius, C. Glotzbach, A. Wölfler, A. Hampel, C. Akal, M. Christl
Spatial patterns of erosion in the central Menderes Massif (Western Turkey) revealed by cosmogenic ^{10}Be - Landscape evolution during active continental extension
Germany, Bremen, 24.-29.09.2017, GeoBremen

K. Hippe, M. Lupker, T. Gordijn, S. Ivy-Ochs, F. Kober, M. Christl, L. Wacker, I. Hajdas, R. Wieler
Unravelling the complex in situ cosmogenic ^{14}C-^{10}Be signature in eroding bedrock surfaces and in river sediment from the Bolivian Altiplano
Austria, Vienna, 27.04.2017, EGU General Assembly

K. Hippe
Unravelling complex landscape evolution with combined in situ cosmogenic ^{14}C-^{10}Be analysis
Germany, Potsdam, 30.10.2017, Colloquium, Institute of Earth and Environmental Science

K. Hippe, M. Lupker, L. Wacker
A second generation ETH in situ cosmogenic ^{14}C extraction line
Canada, Ottawa, 16.08.2017, AMS14

K. Hippe
Constaining landscape evolution with in situ cosmogenic ^{14}C-^{10}Be analysis
Germany, Berlin, 14.12.2017, Seminar

A. Hölzer, M. Gorny, B. Riebe, C. Vockenhuber, C. Walther
^{129}I and ^{127}I Species in Surface Waters form the vicinity of La Hague
Germany, Berlin, 03.-08.9.2017, ICRER 2017

M.L. Huber, S.F. Gallen, M. Lupker, N. Haghipour, M. Christl, A.P. Gajurel
Assessing the origins, timing and transport distances of large exotic boulders in trans-Himalayan rivers
Switzerland, Davos, 17.11.2017, SGM 2017

M.L. Huber, S.F. Gallen, F. Sean, M. Lupker, N. Haghipour, M. Christl, A.P. Gajurel
Assessing the origins, timing and transport distances of large exotic boulders in trans-Himalayan rivers
Switzerland, Davos, 18.11.2017, SGM 2017

E. Huysecom, I. Hajdas, A. Mayor, M.-A. Renold, H.-A. Synal
Protection of the Past—how can science help?
Switzerland, Zurich, 16.-17.11.2017, 14C and the protection of cultural Heritage, ETHZ

H. Kerschner, F. Kober, B. Salcher, M. Christl, C. Schlüchter
The production rate of cosmogenic ^{10}Be at the Koefels rockslide site, Austria
Austria, Vienna, 26.04.2017, EGU General Assembly 2017

S. Ivy-Ochs, J. Braakhekke, G. Monegato, F. Gianotti, G. Forno, K. Hippe, M. Christl, N. Akçar, C. Schlüchter
How well do we really know the timing and extent of glaciers during the Last Glacial Maximum in the Alps?
Austria, Vienna, 27.04.2017, EGU General Assembly

S. Ivy-Ochs, S. Martin, P. Campedel, K. Hippe, C. Vockenhuber, G. Carugati, M. Rigo, D. Pasqual, A. Viganò
Geomorphology and age of large rock avalanches in Trentino (Italy): Castelpietra
Slovenia, Ljubljana, 29.5.-2.6.2017, 4th World Landslide Forum

S. Ivy-Ochs
The dating of rock surfaces with cosmogenic nuclides
Switzerland, Andeer, 10.06.2017, Workshop on Schalenstein in region of Andeer

S. Ivy-Ochs, S. Martin, P. Campedel, K. Hippe, C. Vockenhuber, G. Carugati, M. Rigo, D. Pasqual, A. Viganò
Geomorphology and age of large rock avalanches in Trentino (Italy)
Italy, Padua, 10.-13.07.2017, Annual Meeting Gruppo Italiano di Geologia Strutturale della Società Geologica Italiana

S. Ivy-Ochs
Into and out of the LGM in the Alps
Switzerland, Zurich, 05.12.2017, ERDW Departments Colloqium

B.N. Jones, Z. Siketic, V. Stoytschew, M. Döbeli, K.-U. Miltenberger, P. Pelicon, B. Jencic, J.F. Dias, H. Gnaser, L. Houssiau, A. Simon, R.P. Webb, J. Matsuo
Assessing the molecular imaging capabilities and reproducibility of MeV-SIMS
China, Shanghai, 09.10.2017, IBA 2017 Conference

A. Kaveh Firouz, J. Burg, N. Haghipour, S. Mandal, R. Elyaszadeh, M. Christl
Spatial variability of ^{10}Be-derived erosion rates in Ghezel-Ozan Basin, NW Iran
Switzerland, Davos, 17.11.2017, SGM 2017

J. Kenyon, K.O. Buesseler, N. Casacuberta, M. Castrillejo Iridoy, S. Otosaka, J. Drysdale, S. Pike
Evolution of surface concentrations of ^{90}Sr and ^{137}Cs through 2016 from the Fukushima Dai-ichi nuclear accident
France, Paris, 13.08.2017, Goldschmidt 2017

O. Kronig, J.M. Reitner, M. Christl, S. Ivy-Ochs, H.-A. Synal
Using ^{10}Be to date the stabilisation age of relict rockglaciers
Switzerland, Zürich, 08.03.2017, AMS Seminar

O. Kronig, J.M. Reitner, M. Christl, S. Ivy-Ochs, H.-A. Synal
Using ^{10}Be to date the stabilisation age of relict rockglaciers
Switzerland, Bern, 29.03.2017, Exogene Geology and Quaternary Global Change Seminar

O. Kronig, J.M. Reitner, M. Christl, S. Ivy-Ochs, H.-A. Synal
Using cosmogenic nuclides to date the stabilisation age of relict rockglaciers
Austria, Vienna, 26.04.2017, EGU General Assembly 2017

O. Kronig, J.M. Reitner, M. Christl, S. Ivy-Ochs, H.-A. Synal
Using ^{10}Be exposure dating to determine the stabilization age of relict rockglaciers
Canada, Ottawa, 16.08.2017, AMS14

P.G. Valla, G.E. King, S. Ivy-Ochs, M. Christl, F. Herman
Reconstruction of glacier fluctuations in the Mont-Blanc massif, western Alps: a multi-method approach
Austria, Vienna, 26.04.2017, EGU General Assembly 2017

C. Litty, F. Schlunegger, N. Akçar, P. Lanari, M. Burn, M. Christl
Chronology, patterns and rates of erosion and deposition processes, western Peruvian Andes
France, Paris, 13.08.2017, Goldschmidt 2017

M. Lupker, K. Hippe, L. Wacker
Paired in-situ ^{14}C and ^{10}Be measurements in a Himalayan catchment: residence time or sediment production process tracer?
Austria, Vienna, 27.04.2017, EGU General Assembly

A. Lyså, E. Larsen, N.aki Akçar, J. Anjar, M. Ganerød, R. van der Lelij, C. Vockenhuber
Multiple glaciations at a young volcanic island - Jan Mayen
Sweden, Kristineberg, 22.-26.05.2017, 5th International conference on palaeo-arctic spatial and temporal (past) gateways

R. Delunel, N. Akçar, F. Schlunegger, M. Christl
^{10}Be-inferred paleoerosion history from >10-Ma-old fluvial deposit in northernmost Chile
Austria, Vienna, 23.-28.04.2017, EGU General Assembly

P. Masqué, M. Castrillejo Iridoy, N. Casacuberta, M. Christl, C. Vockenhuber, J. Garcia-Orellana, H.-A. Synal
Anthropogenic ^{129}I and ^{236}U along the GEOTRACES-GA01 transect in the North Atlantic
France, Paris, 13.08.2017, Goldschmidt 2017

S. Maxeiner, H.-A. Synal, M. Christl, L. Wacker, A. Herrmann
Gas flow dynamics in an AMS stripper
Germany, Mainz, 09.03.2017, DPG-17 Conference

S. Maxeiner, H.-A. Synal, M. Christl, M. Suter, A. Müller, C. Vockenhuber
Proof-of-principle of a compact 300 kV multi-isotope AMS facility
Canada, Ottawa, 14.08.2017, AMS14

K. Mettler, O. Fredin, A. Romundset, M. Christl, C. Vockenhuber, N. Akçar
Reconstruction of post-glacial isostatic uplift rates and relative sea level changes in northern Norway with cosmogenic nuclides
Switzerland, Davos, 17.11.2017, SGM 2017

K.-U. Miltenberger, M. Schulte-Borchers, A. Müller, M. George, M. Döbeli, H.-A. Synal
Exploring MeV-SIMS with a capillary microprobe
Germany, Mainz, 07.03.2017, DPG-17Conference

K.-U. Miltenberger, M. Schulte-Borchers, A. Müller, M. George, M. Döbeli, H.-A. Synal
Exploring MeV-SIMS with CHIMP
China, Shanghai, 09.10.2017, IBA 2017 Conference

A. Moran, A., S. Ivy-Ochs, H. Kerschner
Ziwundaschg-^{10}Be dating an Older Dryas cirque glacier moraine in the middle of the Eastern Alps
Austria, Vienna, 27.04.2017, EGU General Assembly

N. Mozafari Amiri, D. Tikhomirov, Ö Sümer. Ç Özkaymak. B. Uzel, S. Yeşilyurt, S. Ivy-Ochs,
C. Vockenhuber, H. Sözbilir, N. Akçar
Destructive earthquakes history of western Anatolia during the last 15 ka
Switzerland, Davos, 17.11.2017 , SGM 2017

E. Opyrchał, J. Zasadni, P. Kłapyta, M. Christl, S. Ivy-Ochs
^{10}Be cosmogenic nuclide chronology of the latest Pleistocene glacial stages in the High Tatra Mountains
Austria, Vienna, 26.04.2017, EGU General Assembly

F. Pagani, E. Stilp, R. Pfenninger, Z. Balogh-Michels, A. Neels, M. Döbeli, A. Remhof, R. Erni, J.L.M. Rupp,
C. Battaglia
Epitaxial thin-film battery anodes and the role of crystallographic orientation on Li-ion conductivity
Switzerland, Geneva, 12.09.2017, Swiss Society of Crystallography Meeting

P.P. Povinec, J. Doric, I. Hajdas, A.J.T. Jull, I. Kontuľ, J. Kaizer, M. Molnár, M. Richtáriková, I. Svetlik,
A. Šivo, E.M. Wild
Radiocarbon dating of wood, charcoal and mortar samples from the Rotunda of St. George (Slovakia)
Hungary, Debrecen, 3.-7.7.2017, 2nd International Radiocarbon in the Environment Conference

D. Brandova, F. Scarciglia, K. Norton, M. Christl, M. Egli
Deciphering Landscape Archives of the Sila Massif by Linking ^{10}Be and $^{239}/^{240}$Pu
France, Paris, 18.08.2017, Goldschmidt 2017

A. Ruppli, D. Brandová, F. Scarciglia, K. Norton, M. Christl, M. Egli
Tracing Landscape Evolution of the Sila Massif using ^{10}Be
Austria, Vienna, 23.-28.04.2017, EGU General Assembly

B. Rea, R. Pellitero, M. Spagnolo, P. Hughes, R. Braithwaite, H. Renssen, S. Ivy-Ochs, A. Ribolini, J. Bakke,
S. Lukas
Atmospheric dynamics over Europe during the Younger Dryas revealed by palaeoglaciers
Austria, Vienna, 23.-28.04.2017, EGU General Assembly

V. Sanial, K.O. Buesseler, M. Charette, S. Nagao, M. Taniguchi, H. Honda, N. Casacuberta
Submarine groundwater discharge: a hidden pathway of Fukushima derived cesium to the ocean off Japan
France, Paris, 13.08.2017, Goldschmidt 2017

M.S. Schwab, N. Haghipour, T.I. Eglinton
Terrestrial organic carbon and plant wax biomarker export from Scottish River Systems
France, Paris, 16.08.2017, Goldschmidt 2017

T. Shinonaga, K. Gückel, M. Christl, J. Tschiersch
Scavenged ^{239}Pu, ^{240}Pu, ^{241}Pu and ^{241}Am from freshly fallen snow on Mt. Zugspitze
Germany, Berlin, 03.-08.09.2017, ICRER

M.H. Simnon, T.M. Dokken, H. Sadatzki, F. Muschitiello, E. Jansen, I. Hajdas
Water column ventilation changes in the eastern Nordic Seas during MIS 3- Insights into the mechanisms of Dansgaard-Oeschger (D/O) cycles
Spain, Zagaroza, 09. -13.5.2017, Pages OSM meeting

A. Sookdeo, L. Wacker, F. Adolphi, J. Beer, U. Büngten, M. Friedrich, G. Helle, B. Kromer, R. Muscheler, D. Nievergelt, M. Pauly, F. Reinig
More wiggles and filling the void: highly resolved temporal 14C dates during the Younger Dyras
Austria, Vienna, 26.04.2017, EGU General Assembly

A. Sookdeo, L. Wacker, F. Adolphi, J. Beer, U. Büngten, M. Friedrich, G. Helle, B. Kromer, R. Muscheler, D. Nievergelt, M. Pauly, F. Reinig
A ^{14}C investigation into the Big Freeze (Younger Dryas)
Canada, Ottawa, 16.08.2017, AMS14

A. Sookdeo, L. Wacker, F. Adolphi, J. Beer, U. Büngten, M. Friedrich, G. Helle, B. Kromer, R. Muscheler, D. Nievergelt, M. Pauly, F. Reinig
Highly resolved radiocarbon dates during the Younger Dryas
Canada, Ottawa, 16.08.2017, AMS14

R. Delunel, F. Schlunegger, N. Akçar, M. Christl
A ^{10}Be-based sediment budget of the Upper Rhône basin, Central Swiss Alps
Austria, Vienna, 23.-28.04.2017, EGU General Assembly

H.-A. Synal
New trends in AMS: What can we expect from future AMS systems?
China, Guilin, 21.11.2017, 7th East Asian AMS Conference

H.-A. Synal
Progress in Accelerator Mass Spectrometry: From research instruments to novel commercial products
China, Taiyuin, 23.11.2017, Seminar at the Chinses Institute for Radiation Protection

H.-A. Synal
Proposal for the 15th AMS conference
Canada, Ottawa, 16.08.2017, AMS14

H.-A. Synal
Improving Designs of Present and Future AMS Systems
Hungary, Debrecen, 10.07.2017, Radiocarbon in the Environment

H.-A. Synal
New Trends in Accelerator Mass Spectrometry (AMS): What can we expect from Future AMS Systems?
Poland, Piaski, 09.06.2017, Mazurian Lake Conference

H.-A. Synal
CO_2 emissions and LIP business travel: Where are we and where do we go?
Switzerland, Zurich, 08.11.2017, AMS Seminar

H.-A. Synal
New Trends in Accelerator Mass Spectrometry (AMS): What can we expect from Future AMS Systems?
Switzerland, Monte Veritas, 07.07.2017, Isotopes 2017

H.-A. Synal
Profession Physics: Live your dreams
Denmark, Copenhagen, 16.05.2017, IPAC 17

H.-A. Synal
Laboratory of Ion Beam Physics: Research & Development activities
Switzerland, Zurich, 02.06.2017, Lunch Seminar D-PHYS

Z. Talip, R. Dressler, J.-C. David, C.Vockenhuber, E. Muller, E. Strub, D. Schumann
Determination of the long-lived radionuclides from proton irradited heavy metal targets
Qatar, Doha, 12.11.2017, 9th International Conference on Isotopes and Expo

M.O. Usman, F. Kirkels, H. Zwart, S. Basu, C. Ponton,T. Blattmann, M. Plötze, N. Haghipour, F. Peterse, M. Lupker, L. Giosan, T. Eglinton
Holocene climate-and anthropogenically-driven mobilization of terrestrial organic matter from the Godavari River Basin, India
France, Paris, 14.08.2017, Goldschmidt 2017

M.O. Usman, F. Kirkels, H. Zwart, S. Basu, C. Ponton,T. Blattmann, M. Plötze, N. Haghipour, F. Peterse, M. Lupker, L. Giosan, T. Eglinton
Large scale Holocene translocation of terrestrial organic carbon from the Godavari River
Italy, Florence, 17.09.2017, International Meeting of Organic Geochemistry

J. Schoonejans, S. Opfergelt, M. Granet, M. Christl, F. Chabaux
Evaluating steady-state soil thickness by coupling uranium series and ^{10}Be cosmogenic radionuclides
Austria, Vienna, 23.-28.4.2017, EGU General Assembly

V. Vanacker, J. Schoonejans, S. Opfergelt, M. Granet, M. Christl, F. Chabaux
Evaluating steady-state soil thickness by coupling uranium series and ^{10}Be cosmogenic radionuclides
Austria, Vienna, 23.04.2017, EGU General Assembly

V. Vanacker, J. Schoonejans, S. Opfergelt, Y. Ameijeiras-Marino, M. Granet, F. Chabaux, M. Christl
Coupling ^{10}Be cosmogenic radionuclides with uranium series to constrain soil production and denudation. Insights from a Mediterranean soilscape
Switzerland, Zurich, 06.12.2017, Seminar

N. Vandermaelen, K. Beerten, V. Vanacker, M. Christl
Constraining exposure age and erosion rates of the Kempen Plateau (Belgium) over the last 1 Ma
Belgium, Louvain-la-Neuve, 24.01.2017, GLOSS workshop

C. Vivo-Vilches, J.M. López-Gutiérrez, M. García-León, C. Vockenhuber, J.L. Leganés
Radiochemical treatment of concrete samples from a nuclear reactor bioshield for ^{41}Ca AMS measurement at CNA Seville (Spain)
Canada, Ottawa, 16.-17.08.2017, AMS14

C. Vivo-Vilches, J.M. López-Gutiérrez, M. García-León, C. Vockenhuber
Factors related to ^{41}K interference on ^{41}Ca AMS measurements and potassium minimization techniques
Canada, Ottawa, 18.08.2017, AMS14

C. Vockenhuber, Marcel Bryner, M. Christl, K.-U. Miltenberger, A. Müller, H.-A. Synal
Optimizing the absorber setup for low energy AMS of ^{10}Be and ^{26}Al
Canada, Ottawa, 19.08.2017, AMS14

C. Vockenhuber, K.-U. Miltenberger, A. Müller, M. Suter, H.-A. Synal
Gas-filled magnet at 6 MV - Isobar suppression for ^{26}Al and ^{36}Cl
Canada, Ottawa, 19.08.2017, AMS14

C. Vockenhuber, M. Christl, K.-U. Miltenberger, A. Müller, K. Hippe, N. Akçar
AMS measurements of ^{26}Al at ETH Zurich
Germany, Mainz, DPG-17 Conference

K.D. von Allmen, J. Ast, X Maeder, M. Döbeli, M. Gindrat, A. Dommann, J. Ramm, A. Neels
BONDCOAT: High efficiency aircraft and land-based gas turbines with a new bond coat concept
Switzerland, Geneva, 12.09.2017, Swiss Society of Crystallography Meeting

K.D. von Allmen, J. Ast, X Maeder, M. Döbeli, M. Gindrat, A. Dommann, J. Ramm, A. Neels
BONDCOAT: High efficiency aircraft and land-basedgas turbines with a new bond coat concept
Switzerland, Fribourg, 01.06.2017, CTI Micro-Nano Event 2017

L. Wacker
Benefits of annually resolved radiocarbon records derived from tree-ring archives
Switzerland, Birmensdorf, 04.04.2017, Tree-ring lecturer

N. Bleicher, U. Büntgen, S. Fahrni, M. Friedrich, A.J. Timothy Jull, B. Kromer, F. Miyake, I. Panyushkina, F.Reinig, A. Sookdeo, W. Tegel
Benefits of annually resolved atmospheric radiocarbon concentrations derived from tree ring records
Hungary, Debrecen, 10.07.2017, Radiocarbon in the Environment

H.-A. Synal
Highest-precision radiocarbon measurements and accurate dating
Canada, Ottawa, 13.08.2017, AMS14

F. Adolphi, N. Bleicher, U. Büntgen, S. Fahrn, M. Friedrich, R. Friedrich, T. Jones, A.J. Timothy Jull, B. Kromer, F. Miyake, R. Mucheler, I. Panyushkina, F. Reinig, A. Sookdeo, H.-A. Synal, W. Tegel, T. Westphal
Towards a new radiocarbon calibration curve based on annually resolved data
Canada, Ottawa, 13.08.2017, IntCal meeting

L. Wacker, N. Bleicher, U. Büntgen, M. Friedrich, R. Friedrich, J.D. Galván, I. Hajdas, A.J. Jull, B. T. Kromer, F. Miyake, D. Nievergelt, F. Reinig, A. Sookdeo, H.-A. Synal, W. Tegel, T. Westphal
Annually resolved atmospheric ^{14}C records reconstructed from tree-rings
Austria, Wien, 26.04.2017, EGU General Assembly

H.-A. Synal
Challenges in Radiocarbon analysis: Small samples and accurate dating
Switzerland, Bern, 14.09.2017, Workshop

L. Wacker
Tree rings and Radiocarbon
Germany, Jena, 25.09.2017, Radiocarbon in the Earth System

C. Walther, S. Schneider, S. Bister, G. Steinhauser, K. Shozugawa, M. Franzmann , L. Hamann, M. Christl, H.-A. Synal, K. Wendt
Plutonium and Uranium release by the FDNPP accident – actinide detection by established and innovative techniques
Japan, Sendai, 09.-14.07.2017, Actinides

A.-M. Wefing, N. Casacuberta, C. Vockenhuber, M. Christl, M.R. van der Loeff
Distribution of ^{129}I in the Fram Strait
France, Paris, 13.08.2017, Goldschmidt 2017

A.-M. Wefing, N. Casacuberta, C. Vockenhuber, M. Christl, M.R. van der Loeff
^{129}I and ^{236}U as tracers of water mass circulation in the Fram Strait
The Netherlands, Texel, 05.12.2017, Seminar

C. Welte, L. Hendriks, L. Wacker, U.M. Hanke, N. Haghipour, T.I. Eglinton, H.-A. Synal
Analysis and data reduction of microscale ^{14}C samples using gas ion source AMS
Canada, Ottawa, 13.08.2017, AMS14

C. Yeman, M. Christl, B. Hattendorf, C. Welte, L. Wacker, A.H. Andrews, H.-A. Synal
Can red snapper live for half a century?
Canada, Ottawa, 18.08.2017, AMS14

C. Yeman, M. Christl, B. Hattendorf, C. Welte, L. Wacker, A.H. Andrews, R. Witbaard, D. Günther, H.-A. Synal
Data reduction of quasi-continuous ^{14}C profiles of carbonates by Laser Ablation AMS
Canada, Ottawa, 13.08.2017, AMS14

C. Yeman, M. Christl, B. Hattendorf, C. Welte, J. Koch, L. Wacker, A.H. Andrews, R. Witbaard, D. Günther, H.-A. Synal
Laser Ablation AMS for online ^{14}C analysis of marine carbonates
Hungary, Debrecen, 03.07.2017, Radiocarbo in the Encironment Conference

C. Yeman, M. Christl, B. Hattendorf, C. Welte, J. Koch, L. Wacker, A.H. Andrews, R. Witbaard, D. Günther, H.-A. Synal
Improvements of the Laser Ablation Interface for Direct ^{14}C- AMS analysis of carbonates
Germany, Mainz, 08.03.2017, DPG-17 Conference

C. Yeman, M. Christl, B. Hattendorf, C. Welte, J. Koch, L. Wacker, A.H. Andrews, R. Witbaard, D. Günther, H.-A. Synal
Laser Ablation Interface coupled to AMS for online ^{14}C analysis of carbonates
Switzerland, Bern, 13.09.2017, Workshop

C. Yeman, M. Christl, B. Hattendorf, C. Welte, J. Koch, L. Wacker, A.H. Andrews, R. Witbaard, D. Günther, H.-A. Synal
Laser Ablation Interface coupled to AMS for online ^{14}C analysis of carbonates
Austria, Vienna, 30.11.2017, Seminar

C. Yeman, M. Christl, B. Hattendorf, C. Welte, J. Koch, L. Wacker, A.H. Andrews, R. Witbaard, D. Günther, H.-A. Synal
Laser ablation interface coupled to AMS for ^{14}C online analysis of carbonates
Canada, Ottawa, 16.08.2017, AMS14

C. Zabcı, T. Sançar, D. Tikhomirov, S. Ivy-Ochs, C. Vockenhuber, A.M. Friedrich, M. Yazıcı, N. Akçar
Cosmogenic ^{36}Cl geochronology of offset terraces along the Ovacık Fault (Malatya-Ovacık Fault Zone, Eastern Turkey): Implications for the intraplate deformation of the Anatolian scholle
Mongolia, Ulaanbaatar, 20.-22.07.2017 , The International Conference on Astronomy & Geophysics in Mongolia

SEMINAR
'CURRENT TOPICS IN ACCELERATOR MASS SPEKTRO-
METRY AND RELATED APPLICATIONS'

Spring semester

03.01.2017
Michael Habicht (University of Zurich), Ancient Egyptian Chronological models and their relation to Radiocarbon dating and Archaeoastronomy

22.02.2017
Rob Witbaard (NIOZ), The ocean quahog: A bivalve with extraordinary longevity as tool to reconstruct climate and environment

01.03.2017
Giovanni Monegato (National Research Council Italy, Geosciences and Earth Resouces), An updated overview of Last Glacial Maximum on the Southern side of the Alps: comparisons and climate considerations

08.03.2017
Naki Akcar (University of Bern), East African Glaciations: First impressions

15.03.2017
Laura Hendriks (ETHZ), Radiocarbon bomb peak dating of paintings

22.03.2017
Núria Casacuberta (ETHZ), Distribution and fate of ^{236}U and ^{129}I in the German GEOTRACES expedition to the Arctic Ocean in 2015

29.03.2017
Olivia Kronig (ETHZ), Using ^{10}Be to date the stablisation age of relict rockglaciers

05.04.2017
Florian Adolphi (Bern/Lund), Assessing and improving the radiocarbon and ice core timescales - insights from cosmogenic radionuclides

12.04.2017
Caroline Heineke (University of Münster), Spatial patterns of erosion during continental extension – A case study from the Menderes Massif, Western Turkey

26.04.2017
Muhammed Usman (ETHZ), Regional-scale Holocene cycling of terrestrial organic matter

03.05.2017
Maxi Castrillejo (Universitat Autònoma de Barcelona), First results of ^{236}U and ^{129}I in the North Atlantic Ocean (GEOVIDE 2014 expedition)

10.05.2017

Biagio Giaccio (CNR, Italy), Paired ^{14}C and ^{40}Ar/^{39}Ar ages of the Campanian Ignimbrite super-eruption (~40 ka): implications for the Late Pleistocene timescale and the evolutionary processes of the Early Upper Palaeolithic in western Eurasia

17.05.2017

Anastasiia Ignatova (ETHZ), An application of radiocarbon measurements for reconstruction the global and regional patterns of carbon burial in marine sediments

24.05.2017

Johannes Lachner, Put a spotlight on ^{36}Cl- and ^{26}AlO-: isobar suppression of cooled ions with photons

31.05.2017

Tallip Zeynep (Paul Scherrer Institut), Determination of the long lived radionuclides from proton-irradiated metal targets

21.06.2017

Maria Belen Roldan (ETHZ), Field work report of the Falàmè Mission in Senegal

12.07.2017

Martin Suter, Charge state distributions and charge changing cross sections and their impact on the performance of AMS

13.07.2017

Marcel Bryner (ETHZ), Low energy accelerator mass spectroscopy of ^{10}Be at TANDY with absorber set-up

Fall semester

13.09.2017

David Skov (University of Aarhus),Detecting climate induced changes of erosion rates using in situ cosmogenic ^{10}Be and ^{14}C

20.09.2017

Hella Wittmann (GFZ Potsdam), Catchment-wide weathering and erosion rates of mafic, ultramafic, and granitic rock from cosmogenic meteoric ^{10}Be/^{9}Be ratios

27.09.2017

Chantal Freymond (ETHZ), Radiocarbon in modern river sediments - a case study on organic carbon transport along the Danube River

04.10.2017

Jixin Qiao (DTU), ^{236}U tracer studies in the Baltic Sea

18.10.2017

Anne-Marie Wefing (ETHZ), Distribution of ^{129}I and ^{236}U in the Fram Strait

25.10.2017
Hannah Gies (ETHZ), Environmental controls on riverine export of organic carbon

01.11.2017
Benjamin Lehmann (University of Lausanne), Combining OSL and ^{10}Be surface exposure dating to constrain ice-extent histories

15.11.2017
Nonhlanhla Dlamini-Stoll (University of Geneva), Burial chronologies: understanding the funerary history of ancient Mali in the Dogon region

22.11.2017
Emma Nilsson (Lund University), Short-term solar variation seen in the ^{10}Be record from the GRIP ice core

29.11.2017
Philip Gautschi (ETHZ), Direct graphitization of CO_2 from whole air samples for radiocarbon analysis

06.12.2017
Veerle Vanacker (University of Leuven), Coupling ^{10}Be cosmogenic radionuclides with uranium series to constrain soil production and denudation. Insights from a Mediterranean soilscape

13.12.2017
Žiga Šmit (University of Ljubljana), The IBA approach to the studies of cultural heritage objects in Slovenia

20.12.2017
Kristina Hippe (ETHZ), The new in situ cosmogenic ^{14}C extraction line at LIP

THESES (INTERNAL)

Term papers/Bachelor

Giacomo Ruggia
Reconstructing the Luteren Valley landscape and palaeoclimate after the LGM
ETH Zurich (Switzerland)

Diploma/Master theses

Franziska Blattmann
Macroscopic Biosignatures of Microbial Activity and its Fossilization Potential
ETH Zurich (Switzerland)

Marius Huber
Assessing the origins, timing and transport distances of large exotic boulders in trans-Himalayan rivers
ETH Zurich (Switzerland)

Nicola Krake
Origin and Distribution of Organic Matter in SW Iberian Margin sediments: Implications for Sea Surface Temperature Proxy Records
ETH Zurich (Switzerland)

Corinne Singeisen
Rock Avalanche Kandersteg: A reconstruction of the landscape evolution in Kander Valley based on field mapping, cosmogenic nuclide dating and runout modelling
ETH Zurich (Switzerland)

Alissa Zuijdgeest
Biogeochemical dynamics of a tropical river-floodplain system
ETH Zurich (Switzerland)

Doctoral theses

Chia-Yu Chen
The Spatial and Temporal Patterns of Erosion Rate in the Southern Central Range of Taiwan
ETH Zurich (Switzerland)

Aline Fluri
Strain Engineering in Thin Film Electrolytes for Solid Oxide Fuel Cells
ETH Zurich (Switzerland)

Chantal Freymond
Transport and evolution of terrestrial organic carbon signals along the Danube River basin
ETH Zurich (Switzerland)

Huan Ma
Microstructure Engineering via Ion-Irradiation in Metallic Thin Films
ETH Zurich (Switzerland)

Tessa Von der Voort
From Mountains to Molecules: Insights into Soil Carbon Dynamics
ETH Zurich (Switzerland)

THESES (EXTERNAL)

Term papers/Bachelor

Amir Sindelar
Altholzeffekte und Sitesampling, Radiokarbondatierungen in Kupferverhüttungskontexten im Oberhalbstein in Graubünden (CH)
University of Zurich (Switzeland)

Diploma/Master theses

Anna-Elina Pasi
Study of environmental impacts of the Fukushima accident- Determination of ^{3}H, ^{90}Sr, ^{129}I and ^{134,137}Cs in surface waters
University of Helsinki/Leibniz Universität Hannover (Finland/Germany)

Ryan Petini
Influences régionales des activités anthropiques sur l'érosion de sols et le transfert de sédiments à travers le système fluvial : Une méta-analyse pour l'Amérique du Sud
Université catholique de Louvain (Belgium)

Simon Pottgiesser
Bestimmung von ^{90}Sr und Plutonium in Umweltproben aus norddeutschen Trinkwassergebiete
Leibniz Universität Hannover (Germany)

Katharina Schoppengerd
The impact of earthquake-induced landsliding on the concentration of ^{10}Be in river sediments, the example of the 2008 Wenchuan earthquake
University of Münster (Germany)

Qing Wu
Instrumental Analysis of Zwischgold
University of Bern (Switzerland

Doctoral theses

Maxi Castrillejo Iridoy
Sources and distribution of artificial radionuclides in the oceans: from Fukushima to the Mediterranean Sea
Universitat Autònoma de Barcelona (Spain)

Loic Fave
Investigation of the thermal conductivity of SiC/SiC cladding before and after irradiation
EPFL (Switzerland)

Stephanie Fernandez
Combining PbTiO$_3$ and SrTiO$_3$ toward 180° ferroelectric domains
University of Geneva
Sandra Jenatsch
Dynamics of Electronic and Ionic Charges in Cyanine Organic Semiconductor Devices
EPFL (Switzerland)

Camille Litty
Conglomerate terraces dating, ^{10}Be, ^{26}Al, erosional mechanisms, surface erosion rate and variations in climate in Peru
University of Bern (Switzerland)

Andrea Madella
A reconstruction of the recent erosional history of the Western Andean Escarpment in Northern Chile with Terrestrial Cosmogenic Nuclides
University of Bern (Switzerland)

Andrew Moran
Alpine glaciers and rock glaciers during the Lateglacial-Holocene transition
University of Innsbruck (Austria)

Nasim Mozafari Amiri
Using cosmogenic ^{36}Cl to determine periods of enhanced seismicity in western Anatolia, Turkey
University of Bern (Switzerland)

Stephanie Schneider
Untersuchung von Umweltproben aus Fukushima in Bezug auf Plutonium und Uran mittels AMS
Leibniz Universität Hannover (Germany)

Grazia Scognamiglio
Optimization of ^{10}Be and ^{26}Al detection with low-energy accelerator mass spectrometry
University of Seville (Spain)

Serdar Yeşilyurt
Late Quaternary Glaciations of Kavuşşahap Mounatins (Eastern –Anatolia) and paleoclimatic implications
Ankara University (Turkey)

COLLABORATIONS

Australia

The Australian National University, Department of Nuclear Physics, Canberra

The University of Western Australia, Oceans Institute, Crawley

Austria

AlpS - Zentrum für Naturgefahren- und Riskomanagement GmbH, Geology and Mass Movements, Innsbruck

Geological Survey of Austria, Sediment Geology, Vienna

University of Innsbruck, Institute of Geography, Geology and Botany, Innsbruck

University of Salzburg, Geography and Geology, Salzburg

University of Vienna, VERA, Faculty of Physics, Vienna

Vienna University of Technology, Institute for Geology, Vienna

Belgium

IMEC, Leuven

Royal Institute for Cultural Heritage, Brussels

Université catholique de Louvain, Earth and Life Institute, Louvain-la-Neuve

Canada

Chalk River Laboratories, Dosimetry Services, Chalk River

China

China Institute for Radiation Protection, Dosimetry Services, Taiyuan city

Peking University, Accelerator Mass Spectrometry Lab., Beijing

University of Science and Technology, Materials Science, Beijing

Croatia

Ruder Boskovic Institute, Ion Beam Interactions, Zagreb

Denmark

Danfysik A/S, Taastrup

Univ. Southern Denmark, Department of Physics, Chemistry and Pharmacy, Odense

Finnland

University of Jyväskylä, Physics Department, Jyväskylä

France

Aix-Marseille University, Collège de France, Aix-en-Provence

Commissariat à l'énergie atomique et aux énergies alternatives, Laboratoire des Sciences du Climat et de l'Environnement (LSCE), Gif-sur-Yvette Cedex

Laboratoire de biogeochimie moléculaire, Strasbourg

LSCE, CNRS-CEA-UVSQ, Gif-sur-Yvette

Université de Savoie, Laboratoire EDYTEM, Le Bourget du Lac

Germany

Alfred Wegener Institute of Polar and Marine Research, Marine Geochemistry, Bremerhaven

Continental GmbH, Liimbach

Deutsches Bergbau Museum, Bochum

GFZ German Research Centre for Geosciences, Earth Surface Geochemistry and Dendrochronology Laboratory, Potsdam

Helmholtz-Zentrum Dresden-Rossendorf, DREAMS, Rossendorf

Helmholtz-Zentrum München, Institut für Strahlenschutz, Neuherberg

Hydroisotop GmbH, Schweitenkirchen

IFM-GEOMAR, Palaeo-Oceanography, Kiel

Leibniz-Institut für Ostseeforschung Warnemünde, Marine Geologie, Rostock

Marum, Micropalaeontology - Paleoceanography and Marine Seimentologie, Bremen

Regierungspräsidium Stuttgart, Landesamt für Denkmalpflege, Esslingen

Reiss-Engelhorn-Museen, Curt-Engelhorn-Zentrum Archäometrie gGmbH, Mannheim

University of Applied Sciences, TH Köln, Technology Arts Sciences , Köln

University of Bochum, Geology, Bochum

University of Cologne, Physics Department and Institute of Geology and Mineralogy, Cologne

University of Hannover, Institute for Radiation Protection and Radioecology, Hannover

University of Heidelberg, Institute of Environmental Physics, Heidelberg

University of Hohenheim, Institute of Botany, Stuttgart

University of Münster, Institute of Geology and Paleontology, Münster

University of Potsdam, Institut für Erd- und Umweltwissenschaften, Potsdam

University of Tübingen, Department of Geosciences, Tübingen

Hungary

Hungarian Academy of Science, Institute of Nuclear Research (ATOMKI), Debrecen

India

Inter-University Accelerator Center, Accelerator Division, New Dehli

Italy

CNR Rome, Institute of Geology, Rome

Geological Survey of the Provincia Autonoma di Trento, Landslide Monitoring, Trento

INGV Istituto Nazionale di Geofisica e Vulcanologia, Sez. Sismologia e Tettonofisica, Rome

University of Bologna, Deptartment Earth Sciences, Bologna

University of Padua, Department of Geosciences, Geology and Geophysics, Padua

University of Pisa, Department of Geology, Pisa

University of Salento, Department of Physics, Lecce

University of Turin, Department of Geology, Turin

Japan

University of Tokai, Department of Marine Biology, Tokai

Liechtenstein

OC Oerlikon AG, Balzers

Oerlikon Surface Solutions AG, Balzers

New Zealand

University of Waikato, Radiocarbon Dating Laboratory, Waikato

Victoria University of Wellington, School of Geography, Environment and Earth Sciences, Wellington

Norway

Norwegian University of Science and Technology, Physical Geography, Trondheim

Rogaland Fylkeskommune, Stavanger

University of Bergen, Department of Earth Science and Biology, Bergen

University of Norway, The Bjerkness Centre for Climate Res., Bergen

Poland

Adam Mickiewicz University, Department of Geology, Poznan

University of Marie Curie Sklodowska, Department of Geography, Lublin

Portugal

Universidade Nova de Lisboa, Departamento de Conservação e Restauro, Lisboa

Romania

Horia Hulubei - National Institute for Physics and Nuclear Engineering, Magurele

Singapore

National University of Singapore, Department of Chemistry, Singapore

Slovakia

Comenius University, Faculty of Mathematics, Physics and Infomatics, Bratislava

Slovenia

Geological Survey of Slovenia, Ljubljana

South Korea

KATRI Korea Apparel Testing and Research Institute, Seoul

Spain

University of Murcia, Department of Plant Biology, Murcia

University of Seville, Physics Department and National Center for Accelerators, Seville

Sweden

Lund University, Department of Earth and Ecosystem Sciences, Lund

University of Uppsala, Angström Institute, Upsalla

Switzerland

ABB Ltd, Lenzburg

ABB USA, Blue Hills

Amt für Kultur Kanton Graubünden, Archäologischer Dienst, Chur

Centre Hospitalier Universitaire Vaudois, Institut de radiophysique, Lausanne

Dendrolabor Wallis, Brig

Empa, Research Groups: Nanoscale Materials Science, Mechanics of Materials and Nanostructures, X-ray Analytics, Energy Conversion, Joining and Corrosion, Thin Films, Dübendorf

ENSI, Brugg

EPFL, Photovoltaics, Lausanne

ETH Zurich, Departments of: Metals Research and Polymers MATL, Trace Element and Micro Analysis , Inorganic Chemistry CHAB, Environmental Physics, Particle Physics, Solid State Physics PHYS, Millimeter Wave ITET, Institute of Geology, Institute of Geochmeistry and Perology ERDW, Zurich

Evatec AG, Trübbach

Fachhochschule Nordwestschweiz, Elektrische Charakterisierung, Brugg

Geneva Fine Art Analysis Sarl, Lancy, Geneva

Gübelin Gem Lab Ltd. (GGL), Luzern

Helmut Fischer AG, Hünenberg

Kanton Bern, Achäologischer Dienst, Bern

Kanton Graubünden, Kantonsarchäologie, Chur

Kanton Solothurn, Kantonsarchäologie, Solothurn

Kanton St. Gallen, Kantonsarchäologie, St. Gallen

Kanton Turgau, Kantonsarchäologie, Frauenfeld

Kanton Zug, Kantonsarchäologie, Zug

Kanton Zürich, Kantonsarchäologie, Dübendorf

Labor für quartäre Hölzer, Affoltern a. Albis

Laboratiore Romand de Dendrochronologie, Moudon

Landesmuseum, Zurich

Nationale Genossenschaft für die Lagerung radioaktiver Abfälle (NAGRA), Wettingen

Office et Musée d'Archéologie Neuchatel, Neuchatel

Paul Scherrer Institut (PSI), Laboratories for Micro and Nanotechnology, for Atmospheric Chemistry, for Radiochemistry and Environmental Chemistry, Materials Group, Hot Laboratories, Villigen

Research Station Agroscope Reckenholz-Tänikon ART, Air Pollution / Climate Group, Zurich

Stadt Zürich, Amt für Städtebau, Zurich

SUPSI, Dipartimento ambiente costruzioni e design (DACD), Lugano

Swiss Federal Institute for Forest, Snow and Landscape Reseach (WSL), Landscape Dynamics, Dendroecology and Soil Sciences, Birmensdorf

Swiss Federal Institute of Aquatic Science and Technology (Eawag), SURF, Dübendorf

Swiss Gemmological Institute, SSEF, Basel

Swiss Institute for Art Research, SIK ISEA, Zurich

University of Basel, Departement Altertumswissenschaften und Institut für Prähistorische und Naturwissenschaftliche Archäologie (IPNA), Basel

University of Bern, Department of Chemie and Biochemistry, Climate and Environmental Physics, Oeschger Center for Climate Research and Institute of Geology, and Geography, Bern

University of Freiburg, Faculty of Environmentat and Natural Resources, Freiburg

University of Geneva, Department of Anthropology and Ecology, Geology and Paleontology, and Quantum Matter Physics, Geneva

University of Lausanne, Department of Geology and Geosciences, Lausanne

University of Zurich, Department of Geography, Institute of Geography, Abteilung Ur- und Frühgeschichte, Zurich

The Netherlands

NIOZ, Coastal Systems Sciences , Texel

Turkey

Dokuz Eylül University, Department of Geological Engineering, Izmir

Istanbul Technical University, Faculty of Mines, Istanbul

Tunceli Üniversitesi, Geology Department, Tunceli

United Kingdom

Brithish Arctic Survey, Cambridge

Cambridge University, Geography, Cambridge

Durham University, Department of Geography, Durham

Queen Mary University of London , School of Geography, London

University of Aberdeen, School of Geosciences, Aberdeen

University of Bristol, School of Chemistry and School of Earth Sciences, Bristol

University of Lancaster, Nuclear Engineering, Lancaster

University of Oxford, Department of Earth Sciences, Oxford

USA

Columbia University, LDEO, New York

Idaho National Laboratory, National and Homeland Security, Idaho Falls

Lamont-Doherty Earth Observatory, Department of Geochemistry, Palisades

NOAA Fischeries, Pacific Islands Fisheries Science Center, Honolulu

University of Utah, Geology and Geophysics, Salt Lake City

Woods Hole Oceanographic Institution, Center for Marine and Environmental Radioactivity, and Marine Chemistry and Geochemistry, Woods Hole

VISITORS AT THE LABORATORY

Silvana Martin
Department of Geosciences, University of Padova, Padova, Italy
31.01.2017 - 10.02.2017 / 11.12.2017 - 15.12.2017

Marc Ostermann
Geology Department, University of Innsbruck, Innsbruck, Austria
13.02.2017 - 17.02.2017 / 27.03.2017 - 31.03.2017

Rob Witbaard
NIOZ, Texel, The Netherlands
22.02.2017 - 23.02.2017

Benjamin Lehmann
University of Lausanne, Lausanne, Switzerland
06.03.2017 - 10.04.2017

Marc Kurz
Woods Hole Oceanographic Institution, Woods Hole, USA
28.03.2017

Franziska Slotta
FB Geowissenschaften, Freie Universität Berlin, Berlin, Germany
03.04.2017 - 13.04.2017 / 09.11.2017 - 19.11.2017

Michael Sigl
Oeschger Center Climate Change Research, University of Bern, Bern, Switzerland
05.04.2017

Caroline Heineke
Geologie und Paläontologie, University of Münster, Münster, Germany
12.04.2017

Stefano Casale
University of Pisa, Pisa, Italy
03.05.2017 - 30.06.2017

Ramón Pellitero Ondicol
University of Aberdeen , Aberdeen, Scotland UK
08.05.2017 - 02.07.2017

Keith Fifield
Australian National University, Canberra, Australia
17.05.2017

Zdravko Siketic
Ruder Boskovic Institute, Division of Experimental Physics, Zagreb, Croatia
02.07.2017 - 07.07.2017

Maria Villa Alfageme
University of Sevilla, Sevilla, Spain
15.07.2017 - 15.09.2017

Nicholas Chu
Woods Hole Oceanographic Institution, Woods Hole, USA
01.02.2017 - 28.02.2017

Hella Wittmann-Oelze
Geochemistry   GFZ, Potsdam, Germany
20.09.2017 - 21.09.2017

Ilya Usoskin
Geophysical Observatory, University of Oulu, Oulu, Finland
12.09.2017

Sami Solanki
MPI für Sonnenforschung, Göttingen, Deutschland
12.09.2017

Florian Adolphi
Oeschger Center Climate Change Research, University of Bern, Bern, Switzerland
12.09.2017 / 21.11.2017 - 22.11.2017

Bernd Kromer
Universität Heidelberg, Heidelberg, Deutscholand
12.09.2017

Daniel Skov
Department of Geoscience, Aarhus University, Aarhus, Denmark
11.09.2017 - 13.09.2017

John Jansen
Department of Geoscience, Aarhus University, Aarhus, Denmark
11.09.2017 - 13.09.2017

Jixin Qiao
Center for Nuclear Technologies, Technical University of Denmark, Roskilde, Denmark
03.10.2017 - 04.10.2017

Irene Schimmelpfennig
CEREGE (Centre Européen de Recherche et d'Enseignement des Géosciences de l'Environnement), Aix-en-Provence, France
16.10.2017 - 17.10.2017

Grant Raisbeck
CSNSM Université Paris Sud, Paris, France
21.11.2017 - 22.11.2017

Raimund Muscheler
Lund University, Lund, Sweden
21.11.2017 - 22.11.2017

Emma Anderberg
Lund University, Lund, Sweden
21.11.2017 - 22.11.2017

Simon Pottgiesser
IRS, University of Hannover, Hannover, Germany
05.12.2017

Veerle Vanacker
Earth and Life Institute , Université catholique de Louvain, Louvain-La-Neuve, Belgium
06.12.2017 - 07.12.2017

Žiga Šmit
Mathematics and Physics, University of Ljubljana, Ljubljana, Slovenia
13.12.2017 - 14.12.2017

TRAINEES AND STUDENTS AT THE LABORATORY

Amir Sindelar
Universität Zurich, Zürich, Switzerland
01.01.2017 - 30.05.2017

Philip Gautschi
ETH Zürich, Zürich, Switzerland
20.02.2017 - 31.08.2017

Tanja Bolliger
Kantonsschule Wettingen, Wettingen, Switzerland
20.02.2017 - 10.03.2017

Reto Schlatter
ETH Zürich, Zürich, Switzerland
07.03.2017 - 31.08.2017

Marcel Bryner
ETH Zürich, Zürich, Switzerland
13.03.2017 - 31.08.2017

Laura Rüegger
Kantonsschule Zofingen, Zofingen, Switzerland
26.06.2017 - 07.07.2017

Julia Bitterli
Kantonsschule Zürich Nord, Zürich, Switzerland
18.07.2017 - 21.07.2017

Stephanie Arnold
ETH Zürich, Zürich, Switzerland
01.10.2017 - 31.03.2017

Nicolas Brehm
ETH Zürich, Zürich, Switzerland
18.10.2017 - 31.04.2018

Clea Thüring
Kantonsschule Olten, Olten, Switzerland
25.09.2017 - 29.09.2017

Catalina Lehmann
Kantonsschule Olten, Olten, Switzerland
25.09.2017 - 29.09.2017